GOAT SCHOOL

Text Copyright © 2011 by Janice Spaulding

All rights reserved.

Goat School™ is a registered trademark owned by Ken and Janice Spaulding, Stony Knolls Farm, Saint Albans, Maine.

ISBN: 978-0-89272-956-2

Library of Congress Cataloging-in-Publication Data

Spaulding, Janice.

  Goat School : a master class in caprine care and cooking / by Janice Spaulding.

    p. cm.

  Includes index.

  ISBN 978-0-89272-956-2

1. Goats. 2. Agriculture--Maine. 3. Farms--Maine. 4. Goat farmers--Maine. 5. Goat milk. 6. Goat cheese. 7. Goat meat. 8. Cooking, American. I. Title.

  SF383.S68 2011

  636.3'9--dc23

                        2011020866

Illustrations by Patrick Corrigan

Design by Chad Hughes

Cover design by Miroslaw Jurek and Ari Meil

Printed in the U.S.A.

5   4   3   2   1

Down East

Books · Magazine · Online

www.downeast.com

Distributed to the trade by National Book Network

GOAT SCHOOL

A Master Class in Caprine Care and Cooking

Down East Books, Camden, ME

JANICE SPAULDING

Table of Contents

7
Introduction

Part I: The Manual

11
Why Raise Goats?

17
Getting Your Goats

35
Breeding Goats & Goat Health

51
Making a Living with Goats

67
Worksheets & Resources

Part II: The Recipes

75
Making Goat Cheese

85
Salads & Small Bites

99
Big Bites & Main Meals

121
Breads

143
Breakfast

151
Cookies

163
Cakes & Other Treats

179
Pickles

191
Acknowledgments

193
Index

To Martha, my "sister" and friend!
Thank you for prodding, joking, and teasing me into doing this book "in my spare time."

Thank you for crying with me when I lost my dog, for laughing with me at the antics of our goats, and just for being there.
You are a true best friend!

Introduction

We started our goat-raising adventure before the invention of the Internet and well before cell phones, back in the days of long-distance telephone charges and monthly periodicals with breeder lists. Imagine trying to do that today! Our first goats were located in Lebanon, New Hampshire. Two angoras gave us our start. One was a doeling and the other a bred doe. We very quickly became hooked. Just like potato chips "you can't have just one." It didn't take long to have a herd of forty pampered, mohair-producing goats. We enjoyed our angoras because we could have the love of our goats and produce a non-destructive product (the mohair).

In 2000, we began to put together a wonderful herd of Boer goats. Even though Boers are a meat goat, our plan was to raise top-quality breeding stock (which we did) so that others could be assured of utilizing locally grown animals for their breeding programs. Maine has a very cold climate, so hearty, locally grown stock is important to ensure a thriving goat population.

As the demand for good-quality Boer goats waned, we recognized a pattern from the past and had already purchased some dairy goats. We now enjoy the fruits of their labor, fresh milk and cheese. This also brings us back to our original goal of raising animals, to harvest a product, in a nondestructive way.

From the very first days of our goat-raising experience we were struck by the lack of information available to beginning breeders. It was apparent that no one wanted to share much information. It became a trial-and-error experience. This has gone on even to the present time. The remark most often heard by us from our "students" always has to do with their surprise in our willingness to share information. Maybe it comes from our own frustration in having to learn the hard way. This to us was always unacceptable. Goats are wonderful, loving, giving creatures and deserve the absolute best care that we can provide. The financial bottom line has never been the focus at Stony Knolls Farm. Our bottom line was, is, and will continue to be the health and well-being of the goats.

It is from that place that we developed Goat School™, where students can come to Saint Albans to learn everything we can teach them in one weekend about goat ownership. It has been so successful that we decided to publish this book as a primer for anyone thinking about getting goats. The following pages are a combination of our original manual for our Goat School™ students and a compilation of a hundred of my ever-popular Goat School™ buffet recipes, plus some of the recipes used to produce what we sell at farmers' markets all summer long.

We hope this book will be the first stop in the wonderful journey that is living with goats.

-Janice

Part I: **The Manual**

Why Raise Goats?

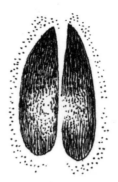

Getting the Goat Bug

Goats are amazing creatures. They are smart, funny, personable, nosy, and did I say smart? They can open any latch, sometimes unlock doors, turn light switches on and off, take things apart, and they love to help! Yes, help. They will help you shovel snow or poop, they will help you put back together what they so nicely took apart, they will untie your shoe laces for you, they love helping take things out of your pockets, they love to help you change that light bulb when you are up on a ladder, but most of all they will help you fall in love with them.

Those beautiful big eyes, their great big hearts, and their sense of humor make you crazy about them. Few people can resist!

And, of course, there are all the good products that they produce (with a little help from you): fiber, meat, dairy, brush clearing, packing; they are also used in the medical research fields. Goats are used to obtain certain serums used in medications (no they do not kill them to do that). Heart medications and pregnancy tests both use blood obtained from goats. Bet you thought that pregnancy tests still used rabbits! Wrong.

But, before you fall totally in love with goats, please be sure you are ready for the commitment. Answer these few questions and find out if you are truly ready for the big step of goat ownership.

Are Goats for Me?

A goat is a big commitment. Before you take the plunge, here are nine important questions to ask yourself. If you don't answer "yes" to most of these, perhaps you should reconsider, and stick with household pets like gerbils or goldfish.

- Can I make myself available twice each day to feed, water, hay, and spend quality time with my goats?

- What is my schedule like? Will that twice-a-day time period fall at the same time each and every day, such as 7 a.m. and 7 p.m.?

- Do I have adequate shelter for these animals?

- Do I have a fenced area that is large enough for the animals to get exercise, and strong enough to keep not only them in, but predators out? Is it tall enough to keep high-jumping goats in? The old saying goes, "If your fence will hold water, it will keep in a goat."

- Can I provide adequate health care for them, either by myself or from a local veterinarian who deals with large animals? (Most dog and cat vets will NOT take care of goats for you!)

- Can I provide good-quality hay for them?

- Is there a grain dealer in my vicinity where I can purchase good, nutritious feeds?

- Am I prepared to teach the goats (remember, you have to have at least two goats, they are herd animals) all of the things it needs to learn, such as walking on a lead, jumping up on a milking stand, coming when called (yes, they learn their names), following you without running off?

- Goats drink enormous amounts of water, am I willing to lug water for them? When my goats are pregnant and due to kid, am I willing to give up going on vacation, going out to dinner, going shopping, going anywhere???

pets

hobby farming

goat-farm business

The Caprine Commitment

{ Getting goats can mean different things to different people. For some, goats are simply useful household pets. Others buy many goats in order to run a for-profit business. Here are a few of the options so you can decide which level of caprine commitment is for you. }

Pets: Brush clearing, and packing, and entertainment are the most common use for pets. They are voracious brush eaters. They will clear out poison ivy, wild raspberries, wild roses, and other thorny overgrown areas. Remember, they will also clean out your garden for you, so be careful!

They are also great pack animals and can learn this fun sport quickly and easily. Let them carry your backpack for you on specially made frames and take them hiking. They love it and you will have a ball.

Hobby Farming: Having a few goats around your farm is great! Milking goats will provide you and your family with fresh milk, from which you can make cheese, soap, and other great goat items! You can freshen (breed) your female goats with a meat buck and produce some of your own meat.

They are great at keeping your pasture weeds to a minimum, allowing other animals to graze contentedly.

Goat-Farm Business: If you are willing to put in the time, a good business can be built around goats! A meat herd will fill not only your freezer, but as long as you use a state or USDA inspected processor, you will be able to sell goat meat. Marketing is a little tough because a lot of people have never tasted goat meat before, however, once they do, they will become steady customers.

Learning to make large volumes of cheese will help you turn a profit at your farm. Farmers' markets, local small gourmet shops, restaurants, health-food stores, and fairs are always looking for purveyors of cheese.

The fiber business is a little more difficult, however, it certainly can be done. Locating the right fiber processor is the most important first step. Finding a raw fleece purchaser can be a daunting task, but, believe me, they are out there, and with the Internet they are only an email or phone call away.

Getting Your Goats

Literally Getting Your Goat

For the first-time goat owner, it's important to know the ins and outs of buying a goat. I highly recommend that you never, ever buy goats at a general livestock auction. Buying your first goats from a good, reputable breeder is tantamount to success. Go to the farm, look around, make sure it's clean. Look at the animals there. Are their eyes clear and bright? Are they nicely filled out, strong, and curious? Are they friendly, free from sores, bumps, and scabs?

When looking to buy milking goats, will the owner let you milk them? If you don't know how to milk, will he/she show you how to do it? NO? Go somewhere else.

Are you buying milk-goat kids? Ask the owner to point out the mom, take a good look at her udder.

All of these things will help you alleviate a lot of heartache, pain, and expense. There is nothing worse than buying an unhealthy goat, or, worse yet, a headstrong, aggressive goat that you can't handle. There is no room on anyone's farm for an unruly goat. Aggressive goats need to take a permanent vacation at "freezer camp."

Goat Whisperer 101
{The Pack, Otherwise Known as the Herd}

How does a herd work, how will this new goat be accepted into my herd, why can't I have only one goat? These questions are ones that deserve some serious thought and consideration!

First of all, goats are very social animals. They are friendly, playful, enjoy interaction with other animals, and love their young. They are inquisitive, sweet, gentle, fiercely, protective, and, of course, always hungry. One goat alone will probably die of boredom, loneliness, or frustration.

Having just one goat as a pet does a great disservice to the animal. A goat that has been raised as a single will have a very difficult time adjusting when put in a herd.

Not too long ago we had to send two goats to "freezer camp" because, no matter what, they could not adjust to living in the herd. They let the other goats push them around, or they would cower in the corner of the barn and refuse to go outside. They had never even seen another goat besides each other! How sad.

In each and every herd there is a queen. Put four unrelated female goats together in a pen and watch! They will head butt, side butt, and make a general nuisance of themselves, but one of the girls will ultimately reign victorious as queen. If you remove that queen, another will step in and take her place in just minutes! Usually there will be a "herd queen" and "first runner up"! Miss first runner up is just chomping at the bit to have her spot in the limelight. Take the queen out of the pen even for a few minutes to hoof trim, and you will see the second-place girl step right up to the plate and take over.

Now keep this in mind when you think about adding some poor little unrelated doeling to your herd! She has no one to keep her safe, no one to cuddle and sleep with, no one to play with, she is missing her mom, and if she was a twin she is really lonesome!

If you are going to add a small kid to your already established herd, please, do it in pairs. Give the little ones a place to hide from the big girls and a place to eat. A grown doe will remove everyone from the feed dish except her own kid; she could care less if there is another kid there with no mom to protect her.

When purchasing a buck, for goodness sake, make sure he has a buddy! A buck needs someone to play with! Someone to head butt with, eat with, and argue with. If a buck has a companion, he will stay even tempered and friendly. CONTINUED ON PAGE 23

Goats of Many Colors

{ Just like dog breeds, goats come in many different sizes, colors, and personalities. Here's a chart to help you decipher which kind of goat or goats is right for you. }

Breed	Function	Defining Characteristics
Alpine	Milk	Large and leggy, with upright ears, come in a variety of colors and markings. Great milkers. Alpines tend to be a little "pushy" and are normally the herd queen if they are in with other goats.
Saanan	Milk	White, with upright ears, very "pretty" and feminine looking. Laid back, large and leggy, terrific milkers, friendly and silly.
Sable	Milk	Variety of colors, with upright ears. A Sable is a colored Saanan. For years, when Saanan breeders got a colored offspring it was disposed of as unusable. Finally, someone figured out that these colored Saanans were pretty terrific goats! A herd book was opened at the registry and the colored Saanans were called Sables. They are great milkers, laid back, friendly, and personable.
Toggenburg	Milk	Beautiful mocha color with either white or cream-colored markings and upright ears. They are docile and gentle, excellent milkers, and produce the cutest babies!
Nubian	Milk	Long, luxurious, pendulous ears, come in a variety of colors and markings, some are even polka dotted! Good milkers, with a lot of butterfat in their milk. Tall, leggy, very laid back, and they talk a lot. These goats will carry on conversations with you!
Oberhasli	Milk	Beautiful mahogany brown with black trim. They can also be all black, but this is not a desirable characteristic in the males. Shorter and rounder than other milk goats, but terrific milkers, good butterfat content, and beautiful, gentle personalities.

Breed	Function	Defining Characteristics
LaMancha	Milk	LaMancha goats are pretty easy to spot in a crowd! Although they do have ear holes, they have little to no outer ear. They are great milkers, have wonderful, sweet, gentle personalities, are large and leggy.
Nigerian Dwarfs	Milk, meat, or pets	Small, short, cute. These half-pint creatures have enormous personalities, talk a lot, give a decent amount of milk for their size, have a high butterfat content, and make great pets!
Pygmy	Great brush goats, good meat goats, fabulous pets, decent milk goats	Short, barrel shaped, with bowed legs. Cute as the dickens! Come in a variety of colors. Their milk has a high amount of butterfat.
Boer	Meat	Large, muscular, white body with a reddish mahogany head and long pendulous ears. They have wonderful, friendly personalities.
Kiko	Meat	Large, white, muscular bodies with longer hair.
Spanish	Meat	Come in a variety of colors and have longish hair. Their ears are often referred to as "airplane ears" because of the way they stick out from their heads.
Myotonic	Meat	Larger sized, with large horizontal ears that do not droop. Myotonics are also called "Tennessee Fainting Goats" or "Stiff Legs."
Angora Goat	Fiber	Medium-size goat with long, luxurious curly white hair. Colored Angoras are becoming popular, but not as well known. Friendly, gentle, fabulous mothers, need to be sheared twice a year.
Cashmere Goat	Fiber and meat	Tall, strong, well-muscled goat with the tendency to produce lots of undercoat called cashmere.
Pygora	Fiber	Cross between an Angora and a Pygmy. Smaller than an Angora, lots of color.

Part I: The Manual

›› Goat Lingo

Now that you're a goat owner, you have to use the correct terminology when speaking to people! Here are a few of the worst offenses:

- A male goat is a buck, NOT a billy. A billy is a $25 goat of unknown origin picked up at a livestock auction.

- A baby goat is a kid.

- A buck's penis is called a pizzle. I can't even tell you how many words I've heard to describe this one!

- Goats kid, they do not calve!

- A female goat is a doe, NOT a nanny goat!

- Freshen means your goat has been bred, and it has kidded, which produces a new milking season. Milk goats must freshen each and every year in order to keep being milked.

- The place that milk comes from on a goat is an udder, NOT a milk bag. For goodness sake, it's like calling a human female's breast a jug, or worse.

- An area set aside for kids to eat is called a creep.

- Goats nurse or suckle; they do not milk their mother. People milk goats, goat kids do not!

Even when adding an adult to your mix, the new goat is often ostracized and lonely. It is always easier and a happier experience for the goats if you add them in pairs.

When adding new adult goats to the herd, you will notice lots of posturing going on. There will be much head butting for the first few days. The "new" goats will have to learn the dynamics of the herd. They have to learn to "worship" the queen.

The best advice for you and your herd, though, is to observe your goats; this is best accomplished at feeding time! The goats basically stay in one place and you can really get a good eye on them. You will be able to notice if one of them is consistently getting pushed aside from the feed pans. Weight loss can result; you may have to take that particular goat aside at feeding time until she is more aggressive at the feed bowl. Morning feeding is the time we check out all our pregnant does during kidding time. It's a good time to give the udders a little squeeze and check their butts for mucous discharge!

At breeding time, a doe in heat can quickly be recognized.

Feeding time is a great time to grab run away goats and squirmy kids for shots, etc!

All in all, the only way you can learn about your herd's dynamics is to watch and learn.

Goat Grub

An entire book could be written about how to properly feed goats, but I'll spare you all the details. Here's a list of the basic food and minerals you'll need to feed your new best friends.

It is difficult when those big caprine eyes are looking at you with the "please feed me, I'm melting away to nothing" look. Don't give in to it. Just make sure they have plenty of hay. That being said, overfeeding of grains and treats can eventually kill your goat. Underfeeding will do the same thing. So how do you reach a happy medium? The ration should be approximately one pound of grain per doe and one to three pounds per buck (depending on size). Also, Don't forget the following specifics:

Let's start with water.

Goats need lots of water, some as much as a gallon a day. It doesn't matter what kind of container you use, KEEP IT CLEAN! Goats will not drink dirty water. With the bucks, you can run into problems with urinary calculi from not ingesting enough water to equalize the phosphorous and calcium they ingest. Angora goats are even fussier than meat and dairy goats are! One little piece of poop and they will avoid that water like the plague.

If your containers start building up algae, you are doing a poor job in your management program. A little bleach mixed with some water and swished around with a brush will keep your containers clean. Goats are not dirty animals, don't treat them that way. If you wouldn't drink out of it, what makes you think your goats will?

Salt blocks are a necessity!

Keep your goats thirsty so that they will drink plenty. Just like humans, the more water they drink, the healthier they will tend to be! A red salt block, also known as a mineral block or brown block, is the best choice because they contain not only salt, but also necessary trace minerals that work well for goats. Red salt blocks have added iron, manganese, copper, and iodine. The goats love them for the salty taste and they are getting additional benefits from the minerals in the blocks!

Hay, hay, hay, and lots of it.

A goat needs approximately 4 percent of its body weight in dry matter per day. That's what they will actually eat, not the other two or three pounds that they spilled on the ground trying to get to that little choice morsel that they see in the very center of the pile. Of course, once it's on the ground, it has gotten stepped on, peed on, and pooped on, so don't expect to ever see your goats bend down and eat it! A well-designed hay feeder will help reduce hay waste.

Which feed to feed?

Now you get the "feed sermon," sheep feed is for sheep, goat feed is for goats, cattle feed is for cattle, etc. If you buy a bag of feed that says it is for sheep or goats, you are buying health problems for your goats.

There are many great goat grains on the market, why jeopardize your goat's health and productivity with a grain formulated for another type of animal? Feeding your goats anything but a formulated goat feed will do them a great disservice.

New Goat-Owner's Checklist
{**So, you just put a deposit on some goats, what now?**}

Things you need before your goats arrive:

- ☐ Feed
- ☐ Hay
- ☐ Shelter
- ☐ Water bucket
- ☐ Feed pan
- ☐ Salt block and holder (preferably a red salt block)
- ☐ Have your fencing up
- ☐ Put a layer of bedding down

What Other Things Will You Need?

{ There is an old Boy Scout adage, "Always be prepared." The same holds true in the goat business. Here is a helpful list of items that will help keep your goats healthy. }

Hoof trimmers Shear Magic Hoof Trimmers are the best.

Dr. Naylor's Hoof N' Heel

Antibiotic Procaine Penicillin G (more commonly called "Pro-Pen-G"). A bottle of LA200 is great to have on hand also!

Ivomec Sheep Drench clear wormer for most types of worms

Safeguard for Goats (white wormer for stomach worms such as tape worms) **(DO NOT give Valbazen to a pregnant doe!)**

Bottle of Listerine (old-fashioned kind, DO NOT get the store brand or any of the "flavored" ones!) A small amount used on a paper towel is worth its weight in gold for cleaning up runny eyes or noses. It's antiseptic, so it makes a great cleanser.

Bottle of Pepto-Bismol (The store brands are sometimes different and can be ineffective for the goat.)

Bar Vac (or other brand that contains both CD/T Perfringens Type C & D Tetanus Toxiod. You will need this for your new kids! Don't let the name scare you, it's just their annual booster shot and it's around $6 for a 50ml bottle!)

Covexin 8 (a great annual vaccine used instead of CD/T. It has the CD/T in it and protects against other Perfringen-type diseases!)

A roll of Vet Wrap (good to have on hand for an emergency.)

Needles and syringes (we recommend 22 gauge with ¾-inch needle. Please do yourself a favor and order Luar lock needles. The Luar Slips can come apart too easily and you will curse them!)

A nylon feeding syringe (made by Cotran — is great to have for administering wormers.)

GOAT TREATS

> Here's a quick goat-treat recipe to whip-up to make your goats happy and healthy! I swear my goats would stand on their heads to get these. Dogs love them, too.

⅛ CUP FRESH GINGER, MINCED

2 TO 3 TABLESPOONS OLIVE OIL

2 CUPS SHREDDED CARROTS

TWO (13-OUNCE) CANS
(or 26 ounces of fresh vegetable broth)

1 CUP GOAT MILK
(or whole cow milk)

1 LARGE CLOVE OF GARLIC, MINCED

1 CUP SUNFLOWER SEEDS

½ CUP GRAPE-NUTS CEREAL

½ CUP CORN MEAL

½ CUP WHOLE-WHEAT FLOUR

2 CUPS SWEET GOAT FEED

2 CUPS RICE KRISPIES CEREAL

1 CUP MOLASSES

1 CUP CORN SYRUP

Preheat the oven to 325 degrees.

In a large skillet, sauté the fresh minced ginger in olive oil until tender, then add the carrots and sauté until tender. Add the vegetable broth and simmer for 5 minutes. Remove from pan and puree in a blender. In a really large bowl, add the puree, milk, garlic, sunflower seeds, grape-nut cereal, cornmeal, flour, goat feed, rice crispy cereal, molasses, and corn syrup.

Mix well; you might have to use your hands to accomplish this.

Add more flour if needed to hold the dough together. Roll into small balls about the size of a walnut, and slightly flatten. Bake for about 45 minutes or until firm and almost crispy.

Part I: The Manual

pointed side out away from the skull

{ eartag placement }

Ear Tagging

We tag meat goats at birth. They grow quickly and if you delay you'll be sure to forget which one was born first. The babies' ears are very thin at this point, and it's easy to insert the tag.

It's a bad thing to be walking around asking, "Who's your mama?" to your baby goats! Use a plastic/paper collar marked appropriately for dairy goats. We mark the dam's name and the kid's date of birth.

Ear tags on a young kid should be put on so that the point of the tag faces the front of the ear, away from the skull. Kids play and they fall and it's a good idea not to have that pointy end facing their skull.

Be very careful about tag placement in relation to the veins that run through the ear. These veins are very obvious, so just take your time. Make sure that when you put the ear tag in, that there is also plenty of room for a tattoo.

If you have several types of goats, i.e., percentages, full bloods, different breeds, you may want to have different colors to designate those goats. You will be able to tell at a glance which goat you are looking at!

You can use a different tag color for different years, sires, family groups, etc.

For our dairy kids, we use pink collars for the girls and blue collars for the boys :).

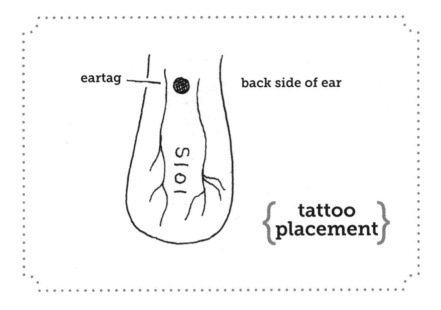

Tattooing

Right ear is your herd prefix

Left ear is the "year letter" and the tag number

Each year is assigned a "year letter" by the goat registries. Any goat born in 2010 should have a "Z" in its ear along with the tag number, for example, Z101. Check with your registry to make sure you are using the right letter for the year according to that particular registry! The dairy registry and meat registry use different letters. Dairy goats born in 2010 have "A" and 2011 is "B"!

The reason for tattooing is tracking. If you should buy a goat with no papers but it has tattoos you can contact goat associations, let them know what the year letter is, the herd prefix, and the tag number, so that if the goat was registered they can help you to obtain its pedigree and registration again! (Be aware that different associations use different letters!)

We use small tattoo letters because we tattoo our kids when they are young. It's easier! Green ink will show up much better than black, especially on a dark colored goat.

The ink is smeared on the ear, the tattoo gun is placed into position, and depressed. After releasing the tattoo gun, more ink is smeared on and then the area is rubbed with baking soda, causing the tattoo to scar the ear. Again, be careful of the veins that run through the ear. Have blood-stop powder on hand just in case!

Caprine Castration
{Or maybe I should title this "Ouch."}

This part of goat ownership makes some men just a little squeamish. Castrating should only be done to those males that will be kept as pets or as companion animals.

There are several methods of castrating, but, here at Stony Knolls Farm, we have decided that banding them is definitely the least stressful and the most kind to the goat.

First thing you will need is an elastrator tool along with the rubber bands. The next is a male goat with these qualifications. He must be at least ten weeks to three months old, his testicles must both have descended and be at least the size of a walnut.

The reasons for waiting until the ten- to twelve-week mark are several. Firstly, do you really want to castrate him? Castrating a male goat that will eventually go to "freezer camp" is NOT a good idea! Castrating stops the production of testosterone, which is responsible for muscle growth. Muscle is meat! By castrating, the growth that the animal will have predominantly fat. For a freezer-destined animal, you want more meat (muscle) and less fat!

Secondly, the young male's urinary tract has got to be developed enough so that the castration won't harm him. If castration is done too early, there is always the possibility of the little guy being prone to urinary calculi due to an underdeveloped urethra, usually resulting in a very painful death.

Angoras grow much slower so it's best to wait until fall shearing. Shear first, then apply the elastic band. This assures no fleece is in the way. A third CD/T shot should be given to an angora buckling at this time to prevent the possibility of tetanus.

Horns

{ To keep them or not to keep them, that is the question. Here are a few horn facts that will give you some perspective. }

❱ **The USA is the only country that removes horns from goats.** It started in the '60's when one state fair decided that horns could injure people peering over the railings at the goats. Ever since, horns have been deemed "dangerous." Not so, but, whatever. We do disbud our baby dairy goats because we sell them, and most people do not want the horns, especially if they are planning on showing the goats.

❱ **Horns are "social" organs; goats use them to re-establish the herd "pecking order," which they do constantly.** Removing the horns does not stop a goat from butting another goat, which is a goat's natural claim to fame. Goats butt with the heavy front plate of their skull, so butting is still going to take place, horns or no horns.

❱ **Horns are a cooling organ** that regulates the temperature of the blood supply to the brain.

❱ **Horns are convenient handles that enable you to control the goat's head** when giving meds or wormers, and also to move a stubborn goat from point A to point B.

> **Horns are useful tools to goats;** they serve not only as back scratchers, but also as a natural working appendage that assists the goat with small daily tasks. (Goat owners may not consider this a "plus factor," since goats are very adept at using their horns to open gates and feed bins, create and enlarge holes in fences, bash down boards in confined areas, etc.)

> **Horns are beautiful!** They are incredible works of art and are interesting to look at and to compare.

Hoof Trimming

{ **Hoof care is one of the most important parts of not only grooming your goat, but also keeping your goat healthy!** }

> **Poorly maintained hoofs can lead to many health issues,** such as hoof rot, sore legs, and eventually severe mutilation of the feet and legs.

> **To observe a healthy hoof, take a good long look at kids' hoofs.** They almost look like they are walking on their tip-toes; the hoof is straight and keeps even pressure on all parts of the foot and leg.

> **Regular trimming is the only way of achieving this goal.** Hoof trimming is fairly simple and the equipment needed is inexpensive and easy to use. A pair of hoof sheers and a hoof pick are all you need!

> **Overgrown or incorrectly trimmed hoofs can be brought back into shape,** but this can be time consuming, and can take months and several trimmings to straighten out the hoof, so it's much better to do it correctly to start with.

> **Start by cleaning the hoof.** Clean out any obvious debris. Take off enough of the wall so that you can see the sole.

> **Trim the hoof walls,** and then even out the sole. Think flat!

> **If you trim a curve into the hoof, the goat will have a difficult time walking.** This is called "rocker hoof," so keep checking to make sure the hoof is flat.

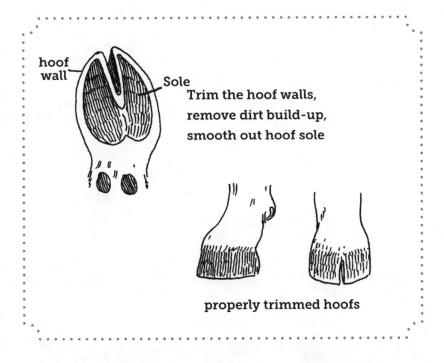

hoof wall
Sole
Trim the hoof walls, remove dirt build-up, smooth out hoof sole

properly trimmed hoofs

❱ **Trimming the heel more than the toe can put uneven pressure on the foot,** causing the goat's pasterns or fetlocks to weaken. Leaving a goat in this condition will eventually cripple it.

❱ **Check between the cloves of the hoof.** Any rash or sores should be addressed immediately! Dr. Naylor's Hoof 'n Heel works terrificly to dry up these problems quickly and easily.

❱ **If you are using a fitting stand, the feet will be much more visible.** Let the goat put the foot down between snips so that you can see how you are progressing. Remember, think flat!

❱ **Back feet are harder to do,** because most goats don't want you messing around back there! Girls nearing the end of pregnancy, milking does, and nursing moms are the worst.

Breeding Goats & Goat Health

Breeding

Oh, the joy of heat, rut, and breeding! Though the topic of breeding goats is large enough for volumes, I'll try to keep it to the most important stuff here.

A doe in heat is usually quite obvious if you have a buck around! She will sometimes moan, groan, yell, flag (tail wag), and make a general nuisance of herself. She will often head-butt the other goats, jump on them like a dog, and be cranky to you, and more often than not have a very red butt! Vaginal discharge is common as well.

A buck in rut is quite comical, actually. They love to display their equipment, pee all over themselves, curl their lip, and make lots of unique noises. The peeing can cause blisters and sores on their noses and very red, irritated-looking eyes. Don't despair, once your does are all bred, rut will be over and the snow and rain will clean away most of the gross, caked-up, yucky stuff on their faces and legs. If you are really obsessive, you can give your boys a bath. Good luck with that.

So, you have a doe in heat and a buck in rut, what now? Check a breeding calendar (see page 71) and make sure that the due date is something you can live with! If so, put them in together and mark down the date! If not, record the heat date and watch for another cycle of 18 to 21 days! Goats are pregnant for approximately 150 days.

A great thing to note here is that if you are planning a vacation, use the chart to make sure your doe won't be giving the farmsitter an unexpected gift! These charts are great for helping you "plan ahead."

Most bucks are extremely gentle to the does. If you have one that isn't, get rid of him! Personality as well as "meatiness" are important factors in your breeding program. Our problems go to "freezer camp" (also known as the "Processor") rather than be sold to an unsuspecting buyer.

The bucks will whisper in the girls' ears, kiss them, and mount them several times. Record this date and watch to see if the doe comes into heat again. This sometimes happens. A doe is in heat anywhere from 8 to 48 hours. But, please note, that every now and then you can have a doe whose heat cycle is even shorter than 8 hours, and if that short period of time occurs during the night, you can miss the cycle altogether!

At the very beginning of the breeding cycle the doe may not be receptive to the buck. If she is in "standing heat," she will go into the pen, squat and pee for him, flag like crazy, and stand for him to mount her.

Some bucks and does are quite fussy about who they breed AND who they are bred with! We put the girls in several times during their cycle, because ovulation can occur anytime during that time period!

If the doe scrunches up during penetration, don't worry. It just means the buck has penetrated her so deeply that he hit her cervix. This is not a problem and it doesn't affect the outcome of the breeding. Also, this doesn't have to happen for her to be bred.

Breeding Techniques

{ There are a couple different types of breeding. Below is a quick summary of each. }

Inbreeding: mating mother to son, father to daughter, or full brother to full sister. This can maximize some really good traits, and also some really bad ones, so be careful what you do here. I do not recommend doing this type of breeding!

Line breeding: breeding half brother to half sister, granddaughter to grandfather, grandson to grandmother. These combinations work pretty well, especially the half brother to half sister! It seems to really maximize the good traits. We have had some amazing kids born from this type of mating.

Crossbreeding, or outbreeding, is crossing two unrelated animals that have characteristics you want to introduce to your flock. This is usually done between different blood lines or different breeds of goats. This type of breeding will produce a hybrid vigor. A good example of this would be to breed a Toggenburg doe with an Alpine buck. A fast-growing, healthy offspring is produced. We have also bred a Boer buck with our dairy does to produce fast-growing meat kids!

We have found over time that good recordkeeping is critical. We refer to our breeding charts from years past over and over again. They contain vital information, which is compiled all in one place! (See page 68 for charts to get you started.)

» Birthing Time:

{ A large canvas bag with handles works great! AND, it's washable! We use this to carry the towels and blow dryer and to weigh the babies. Here's what to include: }

■ **Towels** – bath size, not hand towels, face cloths, or rags! This makes me crazy! I have been to several farms to help deliver kids, and when I ask for a towel I am handed goodness knows what. Rags and small towels so thin you can see through them or so full of holes they look like swiss cheese are inadequate, to say the least!

For goodness sake, buy yourself a gift: six new bath towels, and take the old ones from your bag! These babies need to be warmed fast and easily when it's cold, how is a paper-thin towel with holes in it going to do the job? AND, for crying out loud, wash them right afterward and dry them so that they are ready for the next birth!

■ **An inexpensive blow dryer** – I picked one up at Wal-Mart for $5. It works when the weather is cold and/or damp to get those kids warm and dried in no time!

■ **A baby monitor** – this is a must, and here is the reason why. You paid a lot of money for your goats, you want to make sure that everything goes fine with the labor and delivery, so the monitor gets YOU all set and ready to go! We use a Fisher-Price monitor and it has worked great for many years! The monitor should be set up a few weeks ahead of time so that you get used to the sounds in the barn. Goats moan and groan during the night and if you set up the monitor only for the last day or two you can plan on running out for many false alarms! Once you are used to the normal sounds in the barn, believe me, you will know the out-of-the-ordinary sound of a goat going into labor!

The Kidding Kit!
{**Everything you will need to successfully kid.**}

Disposable gloves — unless you like the feel of yucky stuff!

K-Y Jelly — please use the plain old K-Y Jelly, not scented or flavored! And nothing else works like good old K-Y, believe me! Please don't use vegetable oil, mineral oil, or anything like that! It's not made for this purpose!

7 percent iodine — there has been lots of talk about this being hard to get, but Jeffers still has something that can be used in its place that is just as effective. You need the strength of the 7 percent to dry up and close off the umbilical cord.

Small disposable cups — for dipping navels. We use the little containers that powdered drink mixes comes in; others have collected pill cups from hospitals or nursing homes. They just need to be small enough not to be wasteful and large enough to fit the umbilical cord in!

A little pad of paper and a pencil — pens don't work in the cold. The pad is to note the ear tag number, the weight, gender, and other important information that you are sure to forget when you get in the house!

Your ear tagger and tags — much easier to do it right at birth, and have accurate weight records that match with the proper goat! It's amazing how different the kids look the next morning! Oh, the horror stories I hear! Even to the point that someone didn't know which kid belonged to which mother! See page 28 for more detailed tagging instructions.

Collar bands — these are used to identify dairy kids that would not be ear tagged at birth. They are plastic and easy to write on, so identification is easy.

Bo-Se and syringes — Bo-Se is selenium, and requires a prescription from your vet. Each kid gets ½ cc at birth. This has made a dramatic difference in the health of our kids and especially their strength! It helps to prevent white muscle disease and gives the babies a great start! Use 22-gauge, ¾" Luar lock syringes.

Part I: The Manual

You've Got to Be Kidding

{ **99.9 percent of the time the whole kidding procedure will go smoothly, and it will progress in the following fashion:** }

Goats are like women in their pregnancy. Some get swollen legs, some get really cranky, and some moan and groan and complain. Others go through their pregnancy and you would never know they are pregnant! We had one Angora doe that we didn't realize was even bred until spring shearing; the shearer flipped her over and lo and behold there was an udder forming!

How do you know when your goat is ready to deliver? Watch. The poor girl will get crankier as she gets closer. Some girls produce a lot of mucous. Some of it is stringy and can hang down quite a ways, even dragging on the ground. This is a sign that labor may take place in a short time or within a few days. Recordkeeping comes in really handy at this juncture, more about paperwork later. Our NanC goes 4 or 5 days with a drippy butt, and other goats do not have any mucous at all. Some of them, as soon as I see the mucous string, go into the kidding pen. This is especially true of Angoras.

Watch their udders. You will see changes as their delivery date nears. In some goats the udder will expand greatly (others will expand a few hours before delivery), and when labor is imminent it gets very big, solid feeling, and almost shiny in appearance.

Not all does do this, but one of the best indicators is calling. The doe will walk around looking like she is in a panic, searching for something. She will call out over and over again. Sometimes it's a very soft call, sometimes a gentle talking to her belly, and sometimes a really loud yelling. She is calling to her baby, which hasn't been born yet. This goat needs to be quickly put in her kidding pen. I've found this usually happens in the morning as you are getting ready to feed, as the goats are feeding, or shortly after eating. The girls enjoy eating first, birthing after.

A goat nearing labor will either eat like she'll never see grain again or she will not have any appetite at all. Again, observe your goats ahead of time so that these changes are apparent to you.

Kinds of Kidding

There is a big difference in the labor and delivery of Angoras, dairy goats, and Boers. Angora's seem to have a very quick labor with a delivery that I describe as stop, drop, and roll. The goat stops in her tracks, squats, and pushes out the baby, the baby rolls out, mom cleans it up, and she's done!

Boers have more of a prolonged labor and need to do some pushing before the baby is born. Dairy goats fall somewhere in between. Please be aware that most dairy goats are not good mothers. They have not learned mothering because they were probably taken away at birth and bottle fed. Many of our dairy girls drop and run! The baby is born and they run away from it!

What to Actually Do in Labor

When we hear the sounds of labor beginning, we grab our "kit," and off we go to the barn. You will know when a goat is in labor. Some goats get so cranky it's amazing, and may even try to bite!

Your doe will sometimes call in what starts out as the same sounds you have been hearing and end with a pushing or straining sound. A water bubble might be visible and will usually break. She will get up, lie down, squat, get up, pee, lie down, and so on, so many times it will make you crazy.

We also have heat lamps ready for possible use, which here in Maine is nearly all the time! If you have a close relationship with your doe, she may not want to have her kids without you being around! They can hold back their labor for quite some time.

Once you see that the goat is actually in labor, you will want to put down clean hay in her pen and get your gloves on. DO NOT put your fingers or hands inside the goat unless absolutely necessary! If you have to do this, remember to give your doe a shot of antibiotic as soon as she is done kidding, even if it was just the tip of your finger! If it becomes evident that the doe needs some help, put some K-Y Jelly on your fingers and insert one finger and massage the orifice of the vulva gently from inside. This will usually relax and lubricate enough for the baby to slide out.

If a doe's water breaks and there is no sign of a baby after one hour, you will need to go in and examine. Close your eyes and think about what you should be feeling;

remember two front legs and a head. The usual problem in this case is that the head is turned toward the back end. You will have to push the baby back a little to get the head to face forward. If you should need to pull, and the doe is having contractions, pull with the contraction, in a downward and outward motion.

The usual position should be a nose between two little hoofs. This is, of course, the perfect position and it doesn't always happen. (Don't be alarmed if you see a little tongue hanging out of the kid's mouth! They nearly always are born this way, and it's really quite cute!)

There are always things that can go wrong. The main thing is DON'T PANIC!

If you are seeing two legs that look like back legs, don't worry, the baby can be born this way, too. Quite often if the doe is having twins or triplets, the second baby may be born feet first. Of course, the doe is relaxed and stretched at this point, so the baby will pretty much slide out.

After the kids are born, they need to be cleaned and wiped down quickly. I usually bring the baby around to the front of mom and clean along with her licking. We work together to keep baby warm and get it dried off. If there is more than one kid, make sure both or all of the babies are kept in front of the doe. You certainly don't want her to reject any of her babies if you'll be keeping them with their mom.

Be aware that sometimes kids are bright yellow when they are born. This will be more evident in the Angoras. They look like little yellow chicks! This is normal. It usually happens when a baby is a day or two overdue. The baby's internal organs are beginning to function and the baby passes some of the meconium into the amniotic fluid, thus coloring it and the baby with it.

Once the babies are dried off, fill the small plastic cup with iodine and dip the first kid's umbilical cord and navel area with the iodine. This will cauterize the area and also disinfect it. Simply, place the cup tightly over the umbilical cord stub and the belly button and turn the kid over slightly. The iodine will get all over the kid, don't worry, it comes off! Please throw away the cup and use a new one for the next kid.

About 50 percent of the time, a goat's teat will have a little waxy plug in the end of it, or over the orifice (more common in Angoras). This needs to be removed so the baby can nurse. By milking a small amount from each teat, you will be assured that the teat is free from this plug. If nothing comes out, gently scrape the end of the teat with your fingernail. In really drastic cases, warm cloths will help. Some kids can nurse the plug out, but others can't, so always make sure you have taken this step.

Going back to the umbilical cord, sometimes the cord breaks by itself, other times you must help it break. Do this by shredding it with your fingernail. Never cut it.

We have a scale hanging in our barn most of the year. Once the baby is cleaned up really well, we put it in the large canvas bag (the one we keep all our "delivery" towels in,) hang it from the scale, and weigh it. (Make sure you calibrate your scale to minus out the weight of the bag!)

As for the bath towels, they are big and absorbent. You can wrap a kid up in them and keep them warm while being cleaned up. Also, they are washable. We shake them out really well to get rid of the hay stuck to them. I also try to pick off the larger pieces of mucous. Wash them in hot water with detergent and bleach and you're good to go for the next round. I might mention here that iodine comes out in the wash.

Caring for Mom

So this is the progression that we use: DRY, DIP, TAG, SHOT, WEIGH, then MOM.

Once the babies are all set, this is the point where your doe needs some attention. She has done some hard work and needs a reward. We fill a small bucket with warm water and molasses (1 gallon of water, ¼ to ½ cup of molasses). It gives the poor girl some extra energy, plus most of them love the taste. They are also very thirsty at this stage. If she did not eat at feeding time, I give her some grain. Always, always give clean hay.

During the three or four days that the doe is in her kidding pen, I will give her extra grain. About 1½ times her normal ration. Our does typically get about one pound of 18 percent protein grain per day, so one and one-half pounds would be a good amount. After her seclusion time is over, if the pen is not in use by another goat, I do put mom and babies back in the pen at night and keep them in until after morning feeding time. This is done for two reasons: one, to protect the babies during feeding frenzy, and the other so that I can give a little extra grain to Mom. During the first few days, the babies are nursing a lot and taking a lot from Mom. It's also at this time the babies will imitate mom and start nibbling on grain!

The Afterbirth

The afterbirth (placenta) usually will be delivered in an hour or two. (However, it can take up to 24 hours sometimes!) A placenta isn't considered retained until after 12 hours. We did have a doe that did not deliver it for 2 days, but it certainly wasn't a problem. Try to watch for it. We dispose of it in empty grain bags unless the doe decides to eat it. I know this is gross, but there are all kinds of nutrients and vitamins in the placenta that are good for the doe and help in her healing after birthing. It also contains hormones that trigger her milk production. Some will eat it and some most definitely will not.

DO NOT pull, tug, or in any way try to get the afterbirth out yourself. Let nature take its course.

Brrrr...Cold Kids

If it is cold out and you choose to use goat coats for your babies, make sure to rub the coat on Mom before putting it on. Also, put the coat on the kid in front of Mom! Make sure the babies are quite dry before you do this! If you notice that the baby is damp underneath, remove ASAP! Kids can die of pneumonia really quickly!

Goat Excrements

This is a bit on the gross side, but if you're gonna have goats, you gotta know!

Mom will get a very yucky, crusty area on and around her tail. Once she is finished streaming (getting rid of what is left in her uterus), which can be up to a month, and it's all dried up and cakey, you can trim it off with scissors. Some of it will pull off and parts will just brush off. It is best to clean her up when she is done streaming, especially in fly season!

Now let's address the kids and their poop. The first poop is a black tar-like substance that hopefully Mom will clean up for you! Otherwise, it's difficult to clean up. Warm water and a good butt soak will work nicely to soften and loosen up this gooey substance. I also use baby shampoo if necessary.

Once the meconium (black poop) passes, the next bowel movements will be bright yellow, about the same color as yellow mustard. This is sticky and messy also! Mom will usually clean this up, too, but if she doesn't, you will have to. I emphasize HAVE to. If this yellow poop cakes over the anal opening, it will get hard and make it impossible for the little one to have a bowel movement. This will eventually cause death. Through the years,

I have found this is more of a problem with Angoras than with any other breeds. I think it's more difficult for the mom to clean up all those little curls around the butt area.

Most of the time you can pull off the stuck-on mess, other times it will take a butt soak.

Moms and Kids

Once you've taken care of the essential kidding procedures, for Pete's sake, leave the doe and her new kid alone. DON'T OBSESS! Leave her alone! Just because you don't see the baby nurse doesn't mean the baby isn't nursing. Don't stand there and watch, you'll just make the mom nervous! If she's a first-time mom, and you're concerned that the kid hasn't nursed yet, check the baby after an hour or so. Stick your little finger in the baby's mouth and see if it's warm. If it is, the kid is nursing! If the mom is a seasoned mom, she knows what to do, believe me!

If the kid is yelling, then, for heaven's sake, see if you can attach it to the teat! Once they nurse they'll find it again. But please wait a while. Some moms won't want to nurse until the afterbirth is delivered.

These little relationships are very important. We feel that splitting mom and kids apart causes depression and general unhappiness. Kids that are sent off to new farms without a kid of its own age to bond with are picked on by the new herd, and definitely ostracized.

Leaving two buddies together or sending a mom and her daughter, or neutered son, is so much less stressful! We do what we do for our goats, not for our customers.

A Doe's Secret Code of Honor

Some time ago I read a little article called the "Doe's Secret Code of Honor," author unknown. When I sat down to write this I decided to share with you the Secret Code of Honor at Stony Knolls Farm. All of our girls are in cahoots and had kept this a secret for quite a while, until I finally figured it out! Here is my goats' secret code passed from mother to daughter:

1. If you are ready to kid any day, the honor of all goats is in your hands. Use this time to avenge all of your barnmates. Think about your friends who had to wear silly hats or antlers for Christmas-card photos, posed for calendar pictures, or how about Halloween, did any of your humans dress you up in unthinkable costumes?

2. When you hear the words, "I can't take it any more, she's never going to kid!" wait three more days.

3. When you hear the words, "She's nowhere near ready to kid, we can go to (fill in the blank)," wait until your humans are all cleaned up, dressed up, and ready to get in the car, then give a good scream and start pushing.

4. For every baby monitor, camera, bell, beeper, or whatever your humans are using to keep tabs on your due date, delay delivery by one day. If they are using audio, one good groan per hour will keep them on their toes!

5. You must keep the end-of-pregnancy waiting game interesting! False alarms are mandatory. Little things like looking at your food as if it were gross, digging a little with your hoofs to make a nice nest, and turning your head and talking to your stomach will always get a rise out of your owners!

6 Here is a really important one: Figure out when your owners get ready to go to bed, when they get up, when they shower, and especially when they sit down to a meal. This cannot be allowed! And, don't ever let them use the bathroom in peace. A huge pushing scream (fake one that is) will get them out to the barn in record time. Make sure you look up at them with big innocent eyes at this point!

7 Feeding time, when all of your friends are really hungry is a great time to go into real labor! Really make a fuss, so that only half the feed pans have been filled, water buckets are all empty, the barn is in chaos because everyone is hungry and need to be milked, and are using their loudest voices to let your owners know they are hungry or full of milk. Take your time, relax, and enjoy how crazy you are making your humans!

8 Start off with a screaming push if you realize your owner is half dressed. Its fun to see them running to the barn and trying to get dressed at the same time, especially if it's below zero!

9 Make sure you never look "that far along"! That way you can have your kids outside and really start a scramble! The best time to do this is when your owner is half dressed and he/she tucks your new slimy baby into their jacket only to find that they threw the jacket on over an unbuttoned pajama top. It's really fun!

10 Make the most of your pregnancy. Beg for food every time there is any human in sight. Your barnmates will love you for this one!

11 This one applies whether you are pregnant or not. Only get your head stuck in the hay feeder when the weather is at its worst, or it's nighttime, otherwise it's really not as much fun. As your owner fumbles with the screwdriver, make sure you look up with your big brown eyes and let them know how appreciative you are.

12 When you are in heat, act like a shameless hussy. Drive those boys to distraction. This will drive your humans to distraction, too!

Now remember, girls, this was designed to drive your human owners crazy. It will gently remind him/her how special goats really are, especially when you present them with a beautiful doeling who you can secretly teach the code to future generations!

EVERYTHING YOU EVER WANTED TO
(or not, but you need to)
KNOW ABOUT GOAT HEALTH

PROBLEM	SYMPTOMS	SOLUTION
Worms	Diarrhea, bloated belly, swelling around the jaw area (a problem referred to as Bottle Jaw)	There are various dewormers for different types of worms. White wormers will take care of most stomach worms, and clear wormers will clear up other types.
Lice	Excessive itching	Ivomec Pour-on for Cattle or Guardstar can be used very successfully. Guardstar should be mixed and sprayed on the goat, and Ivomec Pour-on must be placed in a syringe without a needle on it and poured along the spine.
Milk Goiters	Found in babies, a distinct swelling around their jaw line and under their chin	This problem rectifies itself as the baby grows and requires less milk, either from the bottle or from its dam.
Sore Mouth	Crusty, scabby formations all around the mouth and nose area. Can also be spread to a nursing dam's teats if the kid has contracted it	This is another problem that pretty much alleviates itself. If a doe gets it on her teats, and it is a really bad case, you may want to take the kid/kids away and bottle-feed them until she is healed up. By then the babies will probably be over it also. Sore mouth runs about a three-week course, and once a goat has it, they seldom ever get it again.
Pink Eye	Runny, red, sore looking eyes, can progress to a white film covering the eye	Use Terramycin. If the case is especially bad, and a white film is covering the eye, and the goat seems blind, begin dosing with LA-200, 2 ccs twice a day for 3 days.
Runny noses/colds	Slight fever, runny nose, congested, listless	There is a product on the market called Vet RX that works terrific for congestion, and a baby aspirin crushed into some water and given with a small syringe (no needle) will help with the fever. Biggest thing to remember here, folks, is that goats can catch a cold from you!
Diarrhea	Runny poop	Try Pepto-Bismol as a first defense. Vanilla yogurt works great also! Make sure the goat stays hydrated through the episode. Gatorade works well, or Pedialite.

PROBLEM	SYMPTOMS	SOLUTION
Coccidiosis	Out-of-control diarrhea, many times with blood in it	Co-rid is a great medication for this! It's mixed with water and can be given orally to a single kid, or put in the water tank or bucket when more than one goat has the problem.
Shipping Fever	Goat seems like its depressed, listless, sometimes has diarrhea	Give the goat some type of pro-biotic to get its rumen going again. Give only hay and fresh water – NO grain until you can see that the goat seems to feel better.
Hoof Scald	Limping, not walking on the hoof at all	Clean out between the cloves of the hoof with some peroxide and give the hoof a squirt of Dr. Naylor's. Do this twice a day until limping stops.
Urinary Calculi	Bucks or bucklings strain to pee, seem to be in a lot of pain	Treat the goat before you ever run into the problem! Keep salt blocks and plenty of clean water in pens. If you do get the problem, call your vet right away!
Goat Polio	Goat is blind, staggering, head drawn back against its side, very upset	Work very quickly, give the goat 5 ccs of thiamin immediately, followed by 2 ccs of thiamin every four hours until all symptoms disappear. Also give the goat 3 ccs of procaine penicillin twice a day until symptoms disappear.

Soft Hoofs

I will mention right here that this is not an emergency, but to a new goat owner this might be confusing. When a kid is born, they have to travel quite a distance down the birth canal. Their hoofs are usually placed on each side of their head. If the hoofs had their permanent hard toenails on them, they would rip and tear everything apart, including Mom's uterus.

The hoofs on a new-born kid are soft and gummy. Sometimes little pieces of them actually fall off! Don't worry, this is normal and by the time the kid is twenty-four hours old it will have nice solid hoofs!

Blackfly Busters

That's right. Even our goat compadres are bothered by blackflies, mosquitos, and no-seeums. A quick and easy tip to keep bugs from "bugging" your goats:

Mix Avon Skin So Soft with a little Citronella, water, and a squirt of Dawn dish soap. Put in a spray bottle, spray top and sides of the goats. This will help keep the bugs away and the goats will smell wonderful! There are no specific amounts of ingredients, so use your own judgment.

Making a Living with Goats

$

What's Up With Goats? {By Ken Spaulding}

I consider myself a fairly knowledgeable person when it comes to the goat industry. I read just about all the online forums and have friends all over the country in the goat business. I have been raising goats since 1989 and in 2007 sold and transported more than 350 goats from Texas to the Northeast! I consider our operation "successful" by most standards, but still have to ask that same question over and over again: "What's up with goats?"

Where is this industry going? Why are there so few farms raising hundreds of goats? Why does the focus seem to be on no more than 25 or 30 animals? Do you really think you can make money with so few goats?

All it takes is a Google search for meat goats or raising meat goats to bring lots of current information to your fingertips. That wasn't true even just a few years ago, but there is now a great deal of factual information about raising and selling meat goats.

In my opinion, the United States is sorely lacking in both the number of goats available for slaughter as well as the number of dedicated breeders.

Here are a couple of excerpts from a simple Google search (interesting the things that come up). People are so worried about marketing their goats, while here is a quote saying that management is more important than marketing.

While marketing is important, it's just a "small part" of the picture, LSU AgCenter agent Cleve Weisgerber said.

"Management is the key," Weisgerber said. "You must produce a quality product before you put it on the market. Once you get a reputation for producing a quality product, marketing is easy."

"Youngsters who raise goats learn lessons like being responsible," Gryder said. (Howard Gryder, president of the Mid-South Goat Masters organization.) "They have to be responsible to take care of a goat — feed it and give it shots and things it needs to live. Young people who raise goats are young people who grow into responsible adults."

Jack Black, a producer from Arkansas, said he goes to as many meetings as he can so he can learn more about raising goats.

"These are very informative meetings," said Black, who has been raising goats for 20 years. "A lot of important information is available, and by attending these meetings, I can hear about it from the experts."

This last statement gives proof positive as to why events like Stony Knolls Farm's Goat School are so important. People seem to have this "go it alone" philosophy that, when it comes to raising animals for market, does not work! It is important to work together. Not everyone can have herds of 300 or more so that they have enough animals for slaughter, but just 20 breeders with 30 or 40 animals apiece can work cooperatively to have total numbers of from 600 to 800 goats. Simple, isn't it?

There is a BIG move in the goat industry to promote shows and fullblood animals.

There is very little room in that portion of the industry. As our military says, it's the "boots on the ground" that count. The "show" market has only limited potential for profit. Remember, your competing against folks all over the country who have been at it much longer than we have. Only a few will do well, BUT in the meat market even the smallest of breeders stands a good chance of profit.

The largest consumer market for goat meat is in the Northeast, it ranges from Washington, D.C., to Boston, Massachusetts. However, when it comes to goat breeders, I believe, you can count the large goat breeders in this area on one hand and have fingers left over.

Where do you stand in this rapidly growing industry?

Goat Worth

This is getting to be one of the most frequently asked questions I receive. What price can I get for my goats? I think I saw a formula on the Internet (and we all know you can trust EVERYTHING you read online). Something like: Multiply my age times 3. Subtract the day of the month. Divide that by the pounds of grain fed each day and add $263 to that amount. Not a very smart formula, is it? Actually, though, the formula some people use to price their animals is not very much different. Okay, then, how can I price my goats?

This, of course, is also one of the most difficult questions to answer because there are so many variables. E.g., what are my expenses...? What are my goals...? What will I be willing to do to follow up with my customers...? Will I deliver...? Will I supply health certificates...? What type (quality) of stock do I have to sell...? What bloodlines...? How old are the goats that I have to sell...? Have they kidded before...? Are they good moms...? Do they have multiple kids...? Have these goats been purchased at production sales for re-sale...? What did I pay for them...? What did it cost me to bring them to Maine for resale...? What would I be willing to pay for the goats I have to sell...? What is the going price for goats similar to mine at other nearby farms...? (Remember, you don't want to leave too much money on the table and you definitely do not want to UNDERPRICE your goats — that hurts everyone.) These are all valid and critical questions that you need to ask yourself. These are also NOT the only questions, because each farm is unique and each farmer must learn what is specific to their operation.

How does Stony Knolls Farm set prices? Did you really think I would tell? Not everyone gives up their biggest secrets, but I feel that it shouldn't be a secret. We all need to remain competitive, we all need to price our animals fairly based on the quality and breeding of our livestock and, most importantly, we all need to give our customers a good experience. Bottom line is that it's all about the goats and the customers. Take care of both of those and they will take care of you!

Market Mistakes

Over and over again I'm completely astounded by ads that I read in newspapers, magazines, and, of course, *Uncle Henry's* (a New England buying magazine)! Spelling mistakes, incorrect grammar, phrases that don't make any sense, and, worst of all, too much leeway in the pricing.

These are valuable animals that you own, don't short-sell them! I copied this ad from *Uncle Henry's*, see anything wrong?

> "2 Boar bucks, 6mos Fullbloods, Boar Buckling 88% 06-02-05 bottle baby. Handsome boys. No reasonable offer refused, call eves pls, 20min from Bangor or 25min from Newport, B\O call ..."

I found this ad to be a conglomeration of words with no distinct meanings. No reasonable offer refused?? What in the world are they thinking? What's a reasonable offer? $50? $200? What then? BEST OFFER? Come on! Are your goats only worth the best offer?

I have seen so many ads for BOAR goats that it's pathetic! For crying out loud, a boar is a male pig! Seriously, would you purchase a Boer goat for $1,000 from someone who can't spell Boer?

When you are ready to start marketing your kids, please be careful what you write and how you write it!

Caprine Competition

Competition serves a purpose in all facets of business. Look at Walmart, for example. I remember when everyone was saying Walmart was going to put everyone else out of business. Did they? Not really. Now that they have been around for a while, some people won't even go there! I, for one, would rather go just about anywhere else than big box stores! Do they still make money? Yes. Do they serve a purpose? Yes. Does everyone go there? NO! Are other stores cropping up? YES! Look at Target or Kohl's, they are mega successful!

If a goat farmer should move in down the road from you, will he be competition? Not necessarily. If he is selling a different breed of goat he won't compete at all, if he is selling the same breed, will he then be competing for the same customers that you have? Let's look at this for a moment.

This farm suddenly moves in and it has whole bunches of goats that are white with reddish heads. You are nervous that he is going to put you out of business. What should you do? This is really going to put you on your toes, isn't it? Do some research! Is he selling his goats to a meat market or is he selling breeding stock like you do? What makes your farm stand out from his? How does his stock really look up close and personal? Is he friendly, inviting, willing to share information? Is he selling fullbloods with great bloodlines or is he selling percentages? Are his goats big muscled, solid on their feet, and friendly, or are they skittish and mediocre?

This may seem like a silly question, but what does the farm itself look like? Is it clean, neat, and welcoming? What about barns, outbuildings, and storage? One term the insurance industry uses is "pride of ownership," how does your farm measure up in this category?

Is there junk in your yard, do your buildings need painting, is your grass mowed, do you have cobwebs and such in your barn? Does your place look like summer camp for the Munsters family? After all, how difficult is it to put a rag on a broom and take the cobwebs down?

What are you doing to attract potential buyers? Do you have clear, concise advertisements, or are they over wordy with grammatical and spelling errors? Take a good look at your advertising, because this will separate your customers into classes and determine who will answer your ad and who will not! Remember, spelling shows knowledge! Not sure? Look it up in the dictionary! Using your computer? Hit spell check, for goodness sake! Boer is spelled *B-O-E-R*, not *B-O-A-R*! A BOAR is a male pig!

No matter what the size is, think of your farm business as just that, a business. How comfortable are you going into a store that is dusty, dirty, and smells bad? If a store sells beds and mattresses and you are looking for a table and chairs, are you going to buy a bed instead? No, you will go to another store and find that table and chairs!

Goat farms are the same way! If people don't find what they are looking for, they will go to another that has the merchandise they want!

DAIRY ✳ FIBER ✳ MEAT
{Different strokes for different kinds of goats.}

Dairy Goats

Dairy goats are very unique! Some are smart, some are crafty, some are opportunists, and some are lovey. Goat milk can be used in several ways to turn your product into a small-scale business! Goat-milk soap is wonderful, goat milk is highly sought after, and goat cheese is an extraordinarily good seller. Check with your state department of agriculture to find what rules and regulations apply in your state. Some states are "right to farm" states, which means you can sell products on your farm without licensing, other require a rigorous licensing application, inspection, and testing.

Your Milk Is What Your Goats Eat: Feed, hay, and water will also affect the taste. If the goats are allowed to eat browse, this may also change the palatability. Flavor needs to be consistent if you are planning on selling it. When we purchase goat milk, it is frozen for the first three days for soap making, and then tasted again. Browse such as wild garlic, skunk cabbage, and plant life like that can really turn milk nasty in flavor!

Testing Out Teats: If the goat is lactating when you buy her, make sure her teats are not only working, BUT also are comfortable for you to milk! Short teats are great for a milking machine, but can be extremely hard on someone with large hands! And, of course, the opposite, large teats and small hands will cause some uncomfortable problems also.

Milking Machines: We mean a literal one. Do you really want or need one? They are expensive, and can be a lot of work to clean. Some goats don't appreciate the noise they make. And, of course, hand milking makes for a close relationship with your goat.

> ## Goat Milk Sommelier
> Different kinds of goats produce different-tasting milk. Some goats have a milder flavor and lower in milk-fats. This will usually give a higher volume, but if you are interested in cheese also, the amount of finished product will be noticeably less. Other goats' milk has a higher butterfat content, and will produce less milk but more cheese, gallon for gallon.

Dairy Goat Grub

Dairy goats eat a lot. Plain and simple, they eat more than any other types of goats. They should consume one pound of grain for maintenance plus one pound for each three pounds of milk they produce. A gallon of milk weighs 8.6 pounds, so a goat producing a gallon a day should eat approximately four pounds of grain a day.

There is much written on the percentage of protein needed by dairy goats, but we found that an 18 percent sweet feed works terrifically.

> ### Butterfat percentages by breed+tasting note
>
> **Alpine 3.5%** High volume and very mild taste
>
> **LaMancha 3.9%** High volume and mild taste
>
> **Nubian 4.6%** Medium volume with a stronger taste, excellent for cheesemaking
>
> **Oberhasli 3.7%** Medium volume with mild taste — it's been said this milk is the most "cow like" in taste
>
> **Saanen 3.5%** Very high volume and very mild taste
>
> **Sables 3.5%** Very high volume and very mild taste
>
> **Toggenburg 3.3%** Medium volume with an extremely mild taste
>
> **Nigerian Dwarf 6%** Low volume with a very strong taste, also excellent for cheesemaking

Most goats are neat eaters and will consume all of the feed in the pan while they are being milked. That being said, let me tell you about the "Alpine thing." Alpines are notorious slobs. They pick up huge mouthfuls of food, look around, and spill a good part of it all over the place. We solved the mess problem, somewhat, by placing a large bakery tray under the milking-stand head piece to catch the excess, which is then put back in the grain bin. It stays clean and off the floor, so this is a great way of alleviating waste.

Dairy Kids

We take our dairy kids away immediately at birth. We bottle-feed them and eventually, if we so choose, they could use a lamb bar. If you have never heard of a lamb bar before, it is a large pail with several nipples attached that can feed several kids at once. We prefer to bottle-feed them individually to keep them really sociable, and have had terrific luck with this procedure.

Babies eat about every four hours during the day. Most of them are terrific at sleeping through the night, or at least from about 10 p.m. to around 6 a.m.

For the first day or so they will consume approximately 2 to 4 ounces per feeding. By the second day they are eating more! Day 3 finds them up to around 8 ounces at a feeding. You will be amazed at how much these little pigs can take in!

We use normal baby bottles and nipples. Cutting the hole in the nipple just a little larger does the trick. I insert a small pair of scissors and just make a little snip. Works great.

Colostrum, the first milk, MUST be heated in a water bath. Putting colostrum in the microwave will yield a large, solid, yucky, yellow lump. Put some water in a pan and stick the bottle in it. Heat to around 101 degrees. Once the colostrum is out of the milk (usually by the second or third day), you can start heating the bottles in the microwave. I heat one bottle 1 minute 15 seconds. 2 bottles 1 minute 45 second, 3 bottles 2 minutes 30 seconds, and 4 bottles 3 minutes.

HOW TO MILK A GOAT

1. Measure out the proper amount of food for that particular goat.

2. Bring your goat to its milking stand and get it securely in place. Make sure it can reach its food, yet be restrained so it can't back out of the head gate.

3. Clean the teats and udder well.

4. Place the milking bucket under the udder.

5. Form a ring with both thumbs and forefingers around the top of both teats.

6. Squeeze this ring closed and, using the rest of your fingers, massage the rest of the milk down and out the orifice. DO NOT PULL ON THE TEATS!

7. Continue milking until the teats are totally empty and flabby, the udder is soft and empty looking.

8. Clean the teats again. We use a product called Fight-Bac, which causes the orifice to constrict and totally close. You don't want bacteria getting up inside that teat!

Build your own milking stand!

You'll need:
2 front legs 53"
2 back legs 16"
2 long side pieces 42"
2 front and back pieces 19½"
2 head stanchion pieces 40½" (centered at 5¾" and 12¾") from the right side as you face it
feed pan holder 19½" install up 1"
2 pieces for feed pan holder 14"
1 piece front brace behind feed pan 22½"
top brace front legs 19½"
right side 1X3 28½" (angled to fit)
1 3½" X ¼" carriage bolt
1 washer and 2 nuts
28" small but strong chain
1 cup hook
1 screw and washer
1" board or ¾" plywood for decking
2½" drywall screws
29" from floor to bottom of feed pan holder

Putting it Together:

Assemble the basic frame with two 42" long side pieces and two 19½" pieces. The 19½" pieces fit on the inside of the side pieces. That gives you a finished rectangle of 22" X 42". Install back leg pieces flush with the top of the frame. Install the two front legs. Mark off 16"s from the bottom of the leg and line that mark with the top of the frame with the legs on the inside of the frame.

For the feed pan holder, take two 14" side pieces and install one 19½" piece vertically so that it is 1" higher than the side pieces. Measure back 4" from the front and install the front brace flat side up. Install holder up 29" from floor.

At the top of the front legs mount another 19½" piece for a brace. Install one 40½" head stanchion centered 5¾" from the right. Install movable stanchion piece centered 12¾" from the right. This is bolted using one carriage bolt with nut and washer.

1" or ¾" boards make the floor and the cup hook and chain form the latch.

Don't Bite with the Hand that Feeds!

A milking stand should be used for the sole purpose of milking and eating. The goats are happy to jump up on them because they know it's a safe place, they will get food, and they will get the pressure in their udders relieved.

Using a fitting stand for hoof trimming, worming, and trimming is a much better idea because the goat does not get fed on it; you can do what needs to be done without problems. Fitting stands are available online in several different places and are normally made of heavy-duty steel.

A Hairy Problem

Some breeds of dairy goats can be quite hairy. Having a lot of hair on the udder and teats can be problematic because the hair gets pulled while milking. Goats don't really like that and can potentially jump or, worse yet, put a foot in the pail! A pair of inexpensive clippers will take care of the problem.

Drying Off a Goat

Goats that are being milked twice a day will continue to milk until they are manually dried off. Drying off means to stop milking altogether in a way that will not make your goat uncomfortable or put them in danger of getting mastitis.

An often-asked question is "How do I dry my goat off." This is what we do. About six weeks before the projected "stop milking" date, we stop milking in the evening. Just milk in the morning and give the milking girls a little more grain during the one feeding a day. We do this for about a month, then start milking only every other day for a week or two, then every two days for about a week, then stop completely.

We continue to feed on the milking stand year-round so that the girls never forget their order of milking, which stand they belong on, and are used to being handled. We can also keep track of pregnancies and imminent delivery signs. This has worked wonderfully for us!

Fiber Goats

There are two types of fiber goats, cashmere goats and Angora goats. And another that is starting to show up on the horizon is a cross between a Pygmy and an Angora, called Pygoras.

Fiber goats are gorgeous animals, but have a unique care system attached to them! They must be housed and yarded in good clean areas. They are extremely work intensive! Before you decide you want to raise fiber goats, please go and talk to someone who already has them. Do yourself a favor and know what you are getting into. I would also suggest that you know how to spin and work with fiber before starting with an endeavor such as this.

Cashmere goats: All goats produce cashmere under their hair (except Angoras) during the winter. Cashmere goats were bred to produce more of this very valuable fiber. During the spring, this cashmere is shed. It must be combed out of the hair, then sent off to be de-haired because it will be loaded with guard hairs that MUST be removed. It is then cleaned, carded, and often combined with a soft type of wool to make yarn.

Angora goats: Actually grow hair; this hair is more commonly known as mohair. They grow their fleeces at a rate of one inch a month, so must be sheered twice a year, once in the spring and once in the fall. The fleeces can then be sent off to a processor to be skirted, cleaned, washed, carded, and finally made into either spinable fiber, called roving, or yarn for knitting or weaving. Mohair must be combined with another fiber, because anything made from it will "grow." Sweaters or any other garment made from pure mohair will stretch completely out of shape, become long and funny looking.

Shearing is a back-breaking job requiring some expensive equipment. I would encourage you to find a good shearer before you even purchase Angora goats. A lot of sheep shearers will not do goats because of their thin, fragile skin. Goats also require a special type of comb, which sheep shearers usually do not have.

Also, keep in mind that nothing will cut into your profit more than an expensive shearer or having lice in your fleece.

Meat Goats

Meat goats are very easy to keep and maintain. They need feed, browse, and hay, and like any goat, plenty of clean water. These are probably the least work-intensive goat. We found that feeding them in the morning, making sure their hay feeders and water tanks were full, kept them nicely through the day.

Every three months they need worming and hoof trimming. You can't ask for much easier maintenance than that! They can be the most difficult to worm and hoof trim because of their size and muscle. A good, strong fitting stand is a must in this situation!

The Other (Other) White Meat

"If you want to know who eats goat, it's anybody but white people, descendants of Northern Europe," said Susan Schoenian, a sheep and goat specialist with the University of Maryland Extension Service. "Now all the immigrants come from every other part of the world, and they all come from goat-eating parts of the world."

I must at this point disagree with the above statement. We have found that as the importance of health and especially the rates of cholesterol are becoming more important, so is the use of goat meat. Goat meat is lower in cholesterol than chicken or turkey. Many cardiologists are now recommending that beef be replaced by goat meat in the diet. And, the Food Network has worked wonders in the goat-meat business by featuring many, many recipes using this great meat!

1.9 million	**58%**	**74,980**	**$21,480,000**	**25 million**	**50 million**
THE NUMBER OF GOATS RAISED ANNUALLY FOR MEAT IN THE U.S.	THE PERCENTAGE INCREASE IN THE NUMBER OF GOATS RAISED ANNUALLY FROM 1997 TO 2002	THE NUMBER OF FARMS RAISING MEAT GOATS	DOLLAR VALUE OF IMPORTED GOAT MEAT FOR 2003	THE NUMBER OF POUNDS OF GOAT MEAT CONSUMED IN THE U.S. IN 1997	THE NUMBER OF POUNDS OF GOAT MEAT CONSUMED IN THE U.S. IN 2003

Get Your Goat to "Freezer Camp"

Finding a good processor is a must. Federal law states that goat meat cannot cross state lines unless it is USDA inspected; however, meat only needs to be state inspected in order to sell locally. Our processor charges one rate to cut, wrap, and flash-freeze our chevon (the proper name for goat meat). There is an additional charge for sausage casings and spices.

We always make an appointment ahead of time and bring the goat there just before it will be dispatched to eliminate any anguish on the goat's part of being separated from its herd and sometimes its family.

Goat Meat Markets

Restaurants, both high quality and ethnic, will be interested in your goat meat.

Farmers' Markets. Make sure you have the proper licensing before you bring your meats to a market!

Sell meat off your farm with proper advertising.

Some grocery markets may be interested in your meats, especially health-food-type markets.

Goats Are Good!
They can be used in a variety of ways and can be complementary to many existing livestock enterprises.
* Land cleaning
* Fire suppression
* Orchard pruning
* Weed control
* Marketing goat meat to restaurants

Goats get a bad rap:
Many times, the mere mention of the word goat brings an anguished look and a disgusted shake of the head by the listener. Images of uncontrollable animals running around eating everything in sight are vivid in many people's minds.

Goats = Green
With the use of the goats, herbicide and pesticide usage can be virtually eliminated. Unlike a bulldozer, they control brush and weeds without disturbing the existing grass and soil. They also do not leave synthetic chemicals that can run off into lakes and streams or be ingested by a cow or other animal. They do leave behind lots of useable fertilizer that breaks down quickly and easily!

Goats, Cows, and Sheep, Oh My!
Goats are ideal candidates for multi-specific, many-species, rotational grazing. The goats can be rotated in to eliminate most of the undesirable vegetation (from a cow's perspective), and then the cows can come behind them to graze the grass without having to pick through as many weeds. Stocking rate: two goats per cow.

Goats are, by preference, browsers rather than grazers (stomach analysis shows approximately 72 percent browse, 28 percent grasses). They balance their diet with a variety of grasses (including tussocky and woody species discriminated against by cattle and sheep), weeds, shrubs, and the leaves of larger trees that are within reach. Goats can therefore be introduced to many cattle and sheep properties with no or minimal reduction in primary stock numbers. In times of drought, goats will generally prove hardier and more reliable as breeders than other livestock.

Worksheets & Resources

Paperwork

Probably one of the most important tasks when raising goats is the paperwork.

First, the registrations and transfers for full-blooded goats is very important! Fill out the paperwork and get it in to the proper registry. Until you do, you have no proof that the doe you love so much is yours!

Remember that each "type" of goat has its own registry. Some, such as meat goats, have several registries that you can join.

Registering your kids is very important also! Our doelings are registered almost immediately. We wait a short time (a month or two at the most) to see how the bucklings are developing. If they are looking good, the registrations are sent in. If we are not satisfied with their looks, temperament, or gain, they are neutered, if they will qualify to be a pet or companion, or left intact if they are headed for the freezer. Neutered goats cannot be registered.

When you send in registration papers, remember to photocopy them before mailing, just in case! The postal service is notorious for losing mail!

Take the time to contact your local USDA office and get your farm registered. You never know when grants or low-interest loans might be available for your type of farm. You will not be informed unless the USDA knows about you! And, please, if you are registered with them and they send out questionnaires and censuses, take the time to fill them out and return them! Without the information that those forms supply, your county may not be able to apply for grants and funds that might be useful to you in the future!

Health records are incredibly important. Write down each and every thing you do to your goat when they are sick. If you need a vet and he comes and asks what you have done so far, you had best be prepared! He'll want to know what you gave the goat and how much of it! If this isn't in writing, in the case of an emergency, you won't remember!

Keep track of hoof trimming, worming, shots, and any unusual behavior or problems. Make notes about heat cycles, pregnancy, doe's behavior during labor, date kids are born, how many, how healthy, what you did to them, their weight, etc. Goats tend to do the same things year after year. Looking at a goat's records and saying to myself, "I remember now, she had swollen legs prior to delivery," or "She was very vocal during her heat cycle," is a great reminder. I also keep track of breeding date, due date, and actual delivery date. If she was two days off cycle this year, she will probably be two days off cycle next year! NICE TO KNOW! Kids' weights are really quite important, especially bucklings! You want nice, fast-growing, healthy buck kids to carry on the breed.

One thing I have recorded is that Nellie needs me to be sitting down next to her so she can be partially in my lap. But, after all, she's Nellie and she thinks (probably because I tell her) that she's special. She is also the only doe that regularly kids in the middle of the night!

DOE RECORD

Name		Ear Tag #	
Date of Birth			
Number in Birth	Does	Bucks	Number Live at Birth
Birth Difficulty: Natural ☐ Assisted ☐ Complex ☐ Birth Notes:			
Weight			
Breed		**Breeder:** Self ☐ Other:	
Register: Yes ☐ No ☐		Registration Number	
Tattoo	Right Ear	Left Ear	
Raised		Purchase Date	Price Paid
Description/Color of Goat			
Ears Erect ☐ Pendulous ☐ Airplane ☐ Gopher ☐			
Sire Name		Sire Registration #	
Dam Name		Dam Registration #	
Notes:			

BREEDING RECORD

Dates Exposed	Bred To	Kidding Date	Number of Bucks/Does

Date Sold		Price Received	
Sold To			
Date Died		Cause of Death	

BUCK RECORD

Name		Ear Tag #	
Date of Birth			
Number in Birth	Does	Bucks	Number Live at Birth
Birth Difficulty: Natural ☐ Assisted ☐ Complex ☐ Birth Notes:			
Weight			
Breed		**Breeder:** Self ☐ Other	
Register: Yes ☐ No ☐		Registration Number	
Tattoo	Right Ear	Left Ear	
Raised	Purchase Date	Price Paid	
Description/Color of Goat			
Ears Erect ☐ Pendulous ☐ Airplane ☐ Gopher ☐			
Sire Name		Sire Registration #	
Dam Name		Dam Registration #	
Notes:			
Date Sold	Price Received		
Sold To			
Date Died	Cause of Death		

Facts & Dosages

Normal Goat Statistics:

Body Temperature: 101.5 to 103.5 degrees F

Pulse/Heart Rate: 60 to 80 beats per minute

Respiration Rate: 5 to 30 breaths per minute

Puberty: 4 to 12 months

Estrus (Heat) Cycle: 18 to 23 days

Length of Heat: 8 to 48 hours

Gestation: 145 to 155 days

Dosages:

Any Ivermectin Products: 3 times recommended dosage except injectible of 3 ccs per 25 pounds (follow directions for injectible)

Safeguard for Goats: Recommended (dosage on label)

Valbazen: Use dosage indicated on label (do not use on pregnant goats)

Aspirin: ½ to a whole baby aspirin every other day, according to size of goat (use ½ on babies)

CD/T (initial shot & booster): 2 ccs no matter how small or large the goat is

Covexin 8: 5ccs initially and 2-cc booster at 60 days old

Pepto-Bismol: According to weight on label for humans

Penicillin: Use according to label for lbs.

Goat Resources

- Jeffers Livestock, www.jefferslivestock.com, is a great place to purchase your supplies. Its service is excellent and prices are reasonable. We are not affiliated in any way with this company except as satisfied customers.
- For a great kidding chart go oklahomashowgoats.com/Breeders/Gestation.htm
- Caprine Supply caprinesupply.com a great goat supply company!

Goat Publications:
- Dairy Goat Journal dairygoatjournal.com
- Ruminations smallfarmgoat.com
- The Boer Goat abga.org
- United Caprine News unitedcaprinenews.com
- Goat Rancher goatrancher.com
- GoatKeeper goatkeeper.ca

Part II: **Recipes**

Experimenting with food is fun! A lot of these recipes are just that, the fruition of successful experimentation. Some of the recipes are old ones from my grandmother, some are the "same old, same old" recipes with a twist.

There is a section on cheese, and I implore you to discover this first. If you can make your own cheeses you can wow everyone with these great dishes. I do a lot of cooking with buttermilk, and dislike the over-pasteurized stuff they sell at the grocery. At the beginning of every milking season, I start a new batch of buttermilk and keep it going for months (see page 77). It's great to always have it at your fingertips!

When you question the section on pickles and what could they possibly have to do with raising goats, well, consider this: mounds and mounds of spent bedding sprinkled liberally with tons of goat poop. The best compost ever for healthy gardens that produce an over-abundance of each thing planted!

The majority of recipes here are dishes we serve at our Goat School, others are recipes used to produce the breads and many of the other things we sell at farmers' markets all summer long. Most use goat products, but some don't.

I hope you enjoy each and every one of them. Happy cooking!

Making Goat Cheese

30-MINUTE MOZZARELLA

> Making cheese is fun and easy! Mozzarella is one of the first cheeses people experiment with. It's fast, fun, and easy to make. And, best of all, it's really good!

MAKES ABOUT 8 OUNCES (10 TO 11 WITH NUBIAN MILK)

1 GALLON COLD GOAT MILK
(you can use cow milk for this also!)

1½ LEVEL TEASPOONS CITRIC ACID
dissolved in ¼ cup cool water. (I use powdered citric acid, but you can use pure lemon juice.)

¼ TEASPOON RENNET
dissolved in ¼ cup cool water

⅛ TEASPOON LIPASE POWDER
dissolved in ¼ cup cool water (optional)

KOSHER SALT

Check the temperature of your milk before starting this recipe. If it's already 55 degrees, skip the first step.

Heat the milk very slowly until it reaches 55 degrees.

When it reaches 55 degrees, add the citric acid, and the lipase if desired. Mix thoroughly and heat to 88 degrees. It should be just starting to curdle. When the milk reaches 88 degrees, add the rennet.

Stir gently and continue to heat to just over 100 degrees; remove from heat.

The curds should be pulling away from the sides of the pot and the whey should be just about clear. If it still looks a little milky, wait a few minutes before proceeding.

Remove the curds with a slotted spoon and place in a 2-quart microwaveable bowl. Press the curds gently with your hands to remove as much liquid as possible and pour it off.

Microwave for 1 minute on high. (Rubber gloves will help at this point.) Remove, knead a bit, and pour off any remaining liquid. Return once more to the microwave for another 35 seconds, squeeze, and drain.

Sprinkle with salt and start playing with it. Stretch, roll into a ball, stretch, and continue until the cheese is shiny. Salt to taste.

Roll into small balls to eat now, or store in the refrigerator for up to ten days.

TIP

I always use lipase in my mozzarella to enhance its flavor. It really makes a difference.

DO NOT use table salt when making cheese. Always use either kosher salt or cheese salt. The table salt will not only taste very strong, but it can also turn your cheese an odd color.

GOAT BUTTERMILK

Making your own buttermilk is much like making the first batch of yogurt. You must use a good starter, then keep it going!

MAKES 1 QUART

1 QUART FRESH GOAT MILK

2 TABLESPOONS STORE-BOUGHT BUTTERMILK

In a quart jar, add the milk and buttermilk, cover, and shake well. Set the jar on the counter top at room temperature for 10 to 12 hours. No touching, no peeking! Refrigerate. It's done. When you need more buttermilk, start another quart jar with fresh goat milk and add 2 tablespoons of the buttermilk from the previous batch. You should never have to buy buttermilk again!

FETA CHEESE

> I make this cheese with fresh milk while it's still warm.
> It's the right temperature and saves a step!

MAKES 8 TO 10 OUNCES

1 GALLON GOAT MILK

¼ CUP GOAT BUTTERMILK
(page 77)

⅛ TEASPOON LIPASE POWDER
dissolved in ¼ cup water

½ TEASPOON LIQUID RENNET
dissolved in ¼ cup water

KOSHER SALT

If you are using cold milk, warm the milk to 86 degrees.

Add the goat buttermilk and the dissolved lipase to the milk and stir well. Let set to ripen for one hour.

Stir the diluted rennet into the milk for 1 minute. Allow the milk to coagulate for 40 minutes. Cut the curds into 1" cubes with a long knife and let them rest about 10 minutes.

Stir the cubed curds gently and pour into a cheesecloth-lined colander. Let the curds drain, then tie up the cheesecloth and let the liquid completely drain off. The cheese should feel dry.

Slice, then salt each piece on both sides and place in a covered container for 2 days at room temperature.

The cheese can now either be put in a salt-brine solution of 1 gallon of water with 14 ounces of salt added, or marinated. If you choose to use the salt brine, the cheese must be refrigerated.

To marinate the feta you will be covering the cheese with oil. By doing this, it will keep indefinitely as long as it is completely covered with the oil.

Use olive oil (although this will turn the cheese a strange color, and olive oil will congeal in the refrigerator after opening), canola oil, or soybean oil. I have found canola oil to be the best!

Add any herbs and flavorings such as: basil, bay leaves, fennel, garlic, hot peppers, marjoram, onion, oregano, peppercorns (any color), rosemary, sundried tomato, thyme, or any combination of the above.

We use a combination of basil, garlic, and rosemary in ours.

GOAT WHOLE-MILK RICOTTA

> I have to say, at this point, the freshly made ricotta rarely reaches the refrigerator!

MAKES 1 POUND

1 GALLON GOAT MILK

¼ CUP CIDER VINEGAR

3 TABLESPOONS BUTTER, MELTED
(never use margarine in this recipe!)

½ TEASPOON BAKING SODA

In a double boiler, heat the goat milk to 195 degrees. Slowly stir in the vinegar. Do not add more vinegar than what the recipe calls for. If at this point the whey is still milky, heat the milk a bit more, to 205 degrees.

Gently place the hot curds in a cheesecloth-lined colander and let them drain for a minute or two. Then place the hot curds in a bowl and add the butter and baking soda and mix until the butter is all combined.

It's ready to use or eat! You can store it in the refrigerator, covered, for 1 week.

{SOFT CHEESE}
Using Your Imagination!

Soft cheese (chevre) can be frozen using a good quality Ziploc bag and making sure that all the air is out of it. I freeze it unseasoned and season it after it's thawed. Here are some of our favorite flavor ideas, but please, use your imagination and experiment!

CHIVES AND GARLIC
DILL
DILL AND GARLIC
FRENCH ONION
PINEAPPLE WALNUT
RANCH
HORSERADISH

QUESO FRESCO

> This cheese is fantastic in stir fry because it holds its shape!

MAKES 1 POUND

2 GALLONS GOAT MILK

½ CUP GOAT BUTTERMILK
(page 77)

¼ TEASPOON LIQUID RENNET
diluted in ¼ cup cool water

KOSHER SALT

Heat the milk to 90 degrees and add the buttermilk, stirring gently, then let it set for a minute. Add the diluted rennet and stir gently again. Allow this mixture to set for 30 to 45 minutes.

Slice the curds into cubes and let them set for 25 minutes. Pour into a cheesecloth-lined colander and gently stir in the salt.

Line a cheese mold with cheesecloth and place the curds into it. Press at 35 pounds for 6 hours.

Wrap in plastic wrap and refrigerate for up to 2 weeks.

STONY KNOLLS FARM CHEVRE

> Everyone's favorite goat cheese! This can be substituted for cream cheese in many recipes — and it's so good on burgers and baked potatoes.

MAKES 40 TO 50 OUNCES

THREE DROPS OF RENNET

⅓ CUP COOL WATER

5 QUARTS GOAT MILK

½ CUP GOAT BUTTERMILK
(page 77), set aside at room temperature to take off the chill

First things first! Mix the three drops of rennet into the cool water and set aside. You will only be using 3 tablespoons of this mixture, the balance will lose its potency, so it must be disposed of.

Place the goat milk in a double boiler.

Heat the goat milk to 160 degrees. A little warmer won't matter, but no cooler than 160 degrees.

Take the pot of warmed milk and put it into a sink of very cold water. I put ice in the water. Bring the milk back down to 80 degrees as quickly as you can.

Add the buttermilk and stir gently. Then add 3 tablespoons of the diluted rennet. Stir gently again.

Cover the pot and wrap it in towels or blankets so that it can maintain its temperature for as long as possible. Keep this covered and don't touch, don't peek, don't do anything to this for 24 hours!

After 24 hours, unwrap, uncover, and gently pour into a colander lined with cheesecloth. Make sure you are doing this in the sink! After an hour or so, you can close up the cheesecloth and either hang the cheese to drain or leave it in the colander and place it in a baking pan. The whey will continue to run out from the curd. It should drain for another 24 hours.

After 24 hours the cheese is done! You can salt it and use it in place of cream cheese, or you can flavor it and use it as spread or dip! (see page 79)

Because you have "heat treated" the milk, the cheese will stay good for up to 2 weeks, and we've used it way longer than that.

Part II: The Recipes

HARD CHEESE OR BRICK CHEESE

> A great cheddar-like cheese.

1 GALLON GOAT MILK

½ CUP GOAT BUTTERMILK
(page 77)

½ TEASPOON LIQUID RENNET
mixed with ½ cup cool water

1 TEASPOON KOSHER SALT

Place the goat milk in a double boiler and heat to 160 degrees. Put the heated milk in a sink of cold water and bring the temperature down to 86 degrees. Use ice in the cold water to bring the temperature down quickly.

Add the buttermilk and let stand for 2 hours to ripen.

Add the rennet mixture and stir gently for 1 minute. Cover and allow curds to start forming. This takes about 45 minutes. You can peek after 45 minutes!

The curd is ready when you can stick your finger into it and it leaves a hole that fills up with whey.

Take a long knife and cut the curds into 1-inch pieces.

Cover the curds and allow to set for about 30 minutes. After 30 minutes, put the pan of curds into a sinkful of hot water and bring their temperature up to 102 degrees. Hold at 102 degrees for about 30 minutes.

Meanwhile, stir the curds every 5 minutes to keep them from matting together. When ready, the curds will look like scrambled eggs and will hold their shape. If they don't, let them stay at 102 degrees a little longer.

Remove from the sink and allow to set for about 1 hour, stirring every 10 minutes. Line a colander with cheesecloth, and lift the curds out of the whey with a slotted spoon.

Rinse the curds with warm water and then let drain for 25 minutes. Work in the salt.

Fold a clean dishtowel lengthwise several times until it is only about 3 inches wide and pin it together. Leave the curds in the cheesecloth and place them in the dishtowel band.

Using two small bricks, press between two boards for 12 hours.

Remove the cheese from the folded dishcloth and remove the cheesecloth. Put the cheese on a cake rack to air dry.

Turn occasionally, and allow to dry until all surfaces are dry to the touch. Lightly salt outside surfaces to help draw moisture out.

Turn daily for 6 or 7 days and salt lightly if needed. When completely dry, apply melted cheese wax and then age 1 to 6 months.

Store cheese in a dry place of 50 to 60 degrees. Turn occasionally to make sure the moisture inside the cheese is distributed evenly.

the recipes

Salads & Small Bites

CHEVON TACO SALAD

> This is always a big hit at our Goat Schools. Our guests love the fresh taste and the wonderful layers!

SERVES 10 TO 12

2 TABLESPOONS BUTTER

1 POUND GROUND CHEVON

1 PACKET TACO SEASONING

1 BAG TORTILLA CHIPS

1 HEAD ICEBERG LETTUCE
chopped and shredded

2 CUPS SHREDDED CHEESE

4 OR 5 ROMA TOMATOES
chopped

1 GREEN BELL PEPPER
chopped

1 SMALL CAN SLICED BLACK OLIVES

6 GREEN ONIONS
both green and white parts, chopped

ONE (16-OUNCE) BOTTLE CATALINA SALAD DRESSING

Melt the butter in a medium-size frying pan over medium heat. Fry the ground chevon until well done, then add the taco seasoning.

Break up the tortilla chips into bite-size pieces and set aside.

In a large bowl, layer the lettuce, cooked chevon, ¾ of the tortilla chips, cheese, tomatoes, green pepper, black olives, and finally the green onions. Pour the bottle of Catalina dressing over the top of the salad very carefully. Sprinkle with the remaining tortilla chips. Do not mix this salad.

GREENS, GOAT CHEESE, AND VINAIGRETTE

> Fast, easy, impressive!

SERVES 6 TO 8

1 TABLESPOON BUTTER

2 TABLESPOONS CHOPPED WALNUTS

1 PINT FRESH BLACKBERRIES

¼ CUP APPLE CIDER

1 TABLESPOON APPLE CIDER VINEGAR

½ TEASPOON PREPARED DIJON MUSTARD

¼ CUP CANOLA OIL

8 OUNCES FRESH GREENS
rinsed well, drained, and dried

4 OUNCES PLAIN CHEVRE

In a small skillet, melt the butter. Add the walnuts and sauté lightly over medium heat. Set aside to cool.

Mash half of the blackberries with a fork. Add the cider, vinegar, mustard, and oil, and mix well. Then add the other half of the blackberries.

Place the greens in a large bowl and mix well with the vinaigrette. You can do this salad either in a large bowl or on individual salad plates.

Sprinkle the chevre on the salad mixture and garnish it with the walnuts.

Part II: The Recipes

GREENS WITH FETA AND CROUTONS

> This is a fabulously easy salad with lots and lots of huge taste! Fresh spinach or any green mixture works great.

SERVES 8 TO 10

8 OUNCES FRESH GREENS
rinsed well, drained, and dried

1 JAR MARINATED FETA
(page 78)

1 CUP CROUTONS

Place the greens in a large bowl, pour the jar of marinated feta over the top and sprinkle with the croutons. Done!

GOAT RICOTTA AND PASTA SALAD

> This is a great change for the usual humdrum macaroni salad! This recipe lends itself well to the addition of any freshly chopped herbs or raw veggies.

SERVES 4 TO 6

1 POUND PASTA
macaroni, ziti, bowties, etc.

2 CARROTS
chopped

2 STALKS CELERY
chopped

4 OR 5 MEDIUM RADISHES
chopped

1 CUCUMBER
seeded and chopped

2 GREEN ONIONS
both green and white parts, chopped

16 OUNCES GOAT RICOTTA
(page 79)

SALT AND FRESHLY GROUND BLACK PEPPER
to taste

Cook the pasta according to the directions on its box to al dente. Rinse well in cold water and let cool.

Place the pasta in a large bowl and toss it with the carrots, celery, radishes, cucumber, and green onions.

Add the whole batch of freshly made, warm ricotta. Season the mixture with the salt and pepper.

Part II: The Recipes

PICO DE GALLO WITH FETA

> This salad is a huge hit everywhere I take it. I also sell this salad at farmers' markets, and it's a big hit!

SERVES 6

- 2 TABLESPOONS CANOLA OIL
- 2 TABLESPOONS WHITE VINEGAR
- 2 TABLESPOONS SUGAR
- 1 TABLESPOON LIME JUICE
- ¼ TEASPOON KOSHER SALT
- ¼ TEASPOON GROUND CUMIN
- ¼ TEASPOON FRESHLY GROUND BLACK PEPPER
- 16 OUNCES WHOLE KERNEL CORN
  cooked and cooled
- ONE (16-OUNCE CAN) BLACK BEANS
  well drained
- 2 ROMA TOMATOES
  chopped
- ½ MEDIUM-SIZE RED ONION
  chopped
- 3 GARLIC GLOVES
  chopped really finely
- 2 TABLESPOONS CHOPPED FRESH CILANTRO
- 6 TO 8 OUNCES CUBED GOAT FETA
  (page 78)

Mix the canola oil, white vinegar, sugar, lime juice, salt, cumin, and pepper together in a small bowl. Set aside.

Mix the corn, beans, tomatoes, onion, garlic, cilantro, and feta together very gently. Pour the dressing over this mixture and toss again, remember — gentle!

CHEVRE-STUFFED DEVILED EGGS

> These deviled eggs are highly addictive and very few people can eat just one!

MAKES 24

12 HARD-BOILED EGGS

8 OUNCES PLAIN CHEVRE
(page 81)

1 TABLESPOON PREPARED BROWN MUSTARD

¼ CUP MAYONNAISE

2 TEASPOONS CELERY LEAF
either fresh or dried

½ TEASPOON SALT

1 TEASPOON FRESHLY GROUND BLACK PEPPER

PAPRIKA

Peel the eggs and slice them in half lengthwise. Remove all of the yolks.

Place the yolks in a medium-size bowl with the chevre, mustard, mayonnaise, celery leaf, salt, and pepper. Mix well.

Put this mixture in a large plastic bag. With scissors, cut a corner off the bag. Squeeze the mixture out of the small hole into the hollowed egg white. Garnish with a light sprinkle of paprika.

Part II: The Recipes

GOAT CHEESE AND HONEY DIP

> Wildflower, raspberry, blueberry, or cranberry honey works great with this wonderful treat and if you make your own crackers, your guests enjoy this treat even more!

SERVES 8

8 OUNCES PLAIN CHEVRE
(page 81)

GOOD-QUALITY HONEY

CULINARY LAVENDER
finely chopped

CRACKERS
(see recipe on page 141 for homemade crackers!)

Form the chevre into a round ball and make a well in the top of the ball. Fill the well with honey and sprinkle with the chopped lavender.

Serve with crackers.

GOAT CHEESE CROSTINI

> This makes a great, simple appetizer. Use an electric knife — it will slice through the crusty baguette very easily and you will be able to get nice, thin slices!

MAKES 32

1 BAGUETTE
sliced very thin

¼ CUP SLIVERED ALMONDS

1 CUP CHEVRE
(page 81)

1 TEASPOON CHOPPED FRESH ROSEMARY

3 TABLESPOONS HONEY

Place the baguette slices on a baking sheet and toast under the broiler until lightly browned. Be careful not to overcook them.

In a small non-stick frying pan, lightly toast the almonds.

Place the chevre in a small bowl with the rosemary and mix well. Spoon the chevre mixture over the toasted baguette slices, drizzle with honey, and sprinkle with almonds.

GOAT CHEESE STUFFED TOMATOES

> These can be grilled or cooked under the broiler.
> They are delicious both ways.

SERVES 4

4 RIPE, MEDIUM-SIZE TOMATOES

2 OUNCES GOAT FETA
crumbled or finely chopped (page 78)

OR

2 OUNCES QUESO FRESCO
shredded (page 80)

¼ CUP CHOPPED OLIVES
(black olives work the best)

2 TABLESPOONS CHOPPED FRESH PARSLEY

2 TABLESPOONS CHOPPED FRESH BASIL

Cut the tops off the tomatoes and discard. Scoop out and finely chop all of the pulp. Combine the chopped pulp with the goat cheese, olives, parsley, and basil. Stuff this mixture into the scooped-out tomato shells.

If broiling, spray an eight-inch square pan with baking spray and place the stuffed tomatoes in it. Broil them approximately 2 minutes, or until the skin begins to blister and the stuffing is browned.

If grilling, spray the grill rack with baking spray before pre-heating the grill. Carefully place tomatoes on grill and cook until the skin begins to brown.

MOZZARELLA GRILLED MUSHROOMS

> A super-simple yet very elegant appetizer. They go fast!

SERVES 4

12 GRAPE OR CHERRY TOMATOES
halved or quartered, depending on their size

3 OUNCES FRESH GOAT MOZZARELLA BALLS
(page 76)

1 TABLESPOON OLIVE OIL

2 GARLIC CLOVES
minced

¼ TEASPOON SALT

¼ TEASPOON FRESHLY GROUND BLACK PEPPER

4 LARGE PORTOBELLO MUSHROOMS

1 TABLESPOON CHOPPED BASIL

In a medium bowl, combine the tomatoes, mozzarella, olive oil, garlic, salt, and pepper. Toss very lightly.

Spray your grill rack with cooking spray. Grill the mushrooms lightly over medium heat for 6 to 8 minutes on each side, or until slightly tender.

Spoon the mozzarella mixture into each cap, and sprinkle lightly with the basil.

TURKEY AND CHEVRE PINWHEELS

> I made these appetizers for Goat School and they disappeared so fast, that I vowed next time I will at least triple the recipe!

MAKES 30

8 OUNCES CHEVRE
(page 81)

½ CUP MANDARIN ORANGES
drained and chopped

½ TEASPOON CURRY POWDER

THREE (12-INCH) WHOLE-WHEAT TORTILLAS

½ POUND SLICED DELI TURKEY
(smoked turkey is awesome!)

3 CUPS FRESH BABY SPINACH
washed and drained

2 GREEN ONIONS
green and white parts, chopped

Make sure the tortillas are at room temperature before you begin. In a medium bowl, mix the chevre, oranges, and curry powder. Spread ½ cup of the mixture over each tortilla. Place a layer of turkey over the cheese, then a layer of spinach, and finally the green onions. Roll up tightly and wrap each roll in plastic wrap. Refrigerate for about 2 hours until they firm up.

Unwrap the rolls and slice immediately. Each can be cut into 10 slices.

ZUCCHINI MINI QUICHES

These mini quiches are easily reheated in the oven and taste just as good! I often make them ahead of time for church suppers and re-heat them at 350 degrees for about 8 minutes. Warning: Highly addictive!

MAKES 24

1 PIE CRUST (SIDEBAR)
Or you can cheat and use the ready-to-bake pie crust from your supermarket

1 CUP GRATED ZUCCHINI
(I chop it up after to make sure it's really fine. Keep the skin on, but remove seeds.)

½ CUP CHOPPED MUSHROOMS

¼ CUP COOKED BACON
crumbled well

1 CLOVE GARLIC
finely chopped

¼ CUP GREEN ONIONS
green and white parts, chopped

½ CUP GOAT MILK

2 EGGS
beaten

½ CUP SHREDDED CHEDDAR CHEESE

¼ TEASPOON SALT

DASH OF FRESHLY GROUND BLACK PEPPER

Preheat the oven to 375 degrees. Roll out the pie crust into a rectangle. Cut twenty-four 3-inch circles out of the dough. Press the circles into mini cupcake pans. Mix the zucchini, mushrooms, bacon, garlic, green onions, milk, eggs, cheddar cheese, salt, and pepper together in a medium-size bowl. Fill each dough lined muffin cup with the filling.

Bake for 15 to 18 minutes, or until light brown and puffed. Cool in the pans. Remove carefully. Serve these heavenly tidbits warm.

{SIMPLE PIE CRUST}

This pie crust can be used in most any recipe requiring crust for a two crust pie.

MAKES 1 PIE CRUST

2 CUPS FLOUR
2 TEASPOONS SALT
¾ CUPS OF SHORTENING
4 TO 5 TABLESPOONS ICE WATER

Mix the flour and salt together in a medium sized bowl. Add the shortening and blend until fine. Add the water one tablespoon at a time, mixing well after each addition. When the dough holds together well, roll out on a floured surface.

the recipes

|97|

Part II: The Recipes

Big Bites & Main Meals

CABBAGE ROLLS FOR LAZY COOKS

> We quadruple this recipe for Goat School! It's always a favorite.

SERVES 4 TO 6

1 LARGE CABBAGE
chopped into bite-size pieces

2 MEDIUM SWEET ONIONS
chopped

1 POUND GROUND CHEVON
(sweet chevon sausage works great also)

SALT AND FRESHLY GROUND BLACK PEPPER
to taste

1 CUP UNCOOKED LONG-GRAIN RICE

THREE (14.5-OUNCE) CANS TOMATO SAUCE

Preheat the oven to 350 degrees.

Spread the chopped cabbage into a well greased 9 x 11 inch pan or a pan similar in size. In a large frying pan, cook the onion over medium heat until it's translucent, and add the ground sausage or loose sweet sausage and brown completely. Sprinkle with salt and pepper to taste, and add the uncooked rice, tomato sauce, and 1 cup of water. Simmer these ingredients for 10 to 15 minutes.

Pour this mixture over the cabbage, but don't stir it!

Bake for 2 hours.

CHEVON AND TATER TOTS DELIGHT

> This is a quick, easy recipe that can be put together from pantry items. I am asked for this recipe more than any other that we serve at Goat School.

SERVES 6 TO 8

2 POUNDS GROUND CHEVON
or loose sweet goat sausage

2 LARGE SWEET ONIONS
chopped

2 BAGS FROZEN TATER TOTS

TWO (26-OUNCE) CANS CREAM MUSHROOM SOUP

1 CUP PARMESAN CHEESE

¼ CUP SHREDDED GOAT MOZZARELLA
(page 76)

Preheat the oven to 350 degrees.

In a large frying pan over medium heat, brown the chevon or sausage with the onions. Place the meat in a well-greased, deep 9 x 13 inch pan. Heat the cream of mushroom soup so that it's easier to pour. Then place the Tater Tots on top of the meat and pour the mushroom soup all over it. Sprinkle with the parmesan cheese and then finally the shredded mozzarella.

We sometimes sprinkle this casserole with sliced black olives also.

Bake for 45 minutes, until the cheese is browned and bubbly.

Part II: The Recipes

CHEVON BURGERS

> These are delicious when served on sandwich-size flat bread for the rolls and garnished with mesclun mix, sliced tomatoes, spread with either fresh chevre or slices of goat mozzarella, then place a big dollop of the roasted red pepper aioli on them.

MAKES 6

1 TABLESPOON OIL
vegetable or olive oil

1 SMALL ONION
minced very fine

1 POUND GROUND CHEVON

1 TEASPOON GRANULATED GARLIC

1 TEASPOON CUMIN

1 TABLESPOON CHOPPED PARSLEY
either fresh or dried

¼ TEASPOON FRESHLY GROUND BLACK PEPPER

1 TEASPOON SALT

⅛ CUP PREPARED BROWN MUSTARD

1 TABLESPOON WORCESTERSHIRE SAUCE

In a large skillet over low heat, add the oil and sauté the onions and cook until lightly caramelized. To the chevon, add the onions, garlic, cumin, parsley, pepper, salt, mustard, and Worcestershire sauce and mix well. Refrigerate this mixture overnight to allow the flavors to develop.

Shape the meat into 6 patties and grill to desired temperature. This can be served rare, medium, or well done!

{ROASTED RED PEPPER AIOLI}

A delicious addition to chevon burgers — or any sandwich

MAKES ¾ TO 1 CUP

1 RED PEPPER

1 SMALL ONION
minced

2 TABLESPOONS BUTTER

1 TABLESPOON MINCED GARLIC

1 TEASPOON SALT

¼ TEASPOON FRESHLY GROUND BLACK PEPPER

½ CUP OLIVE OIL MAYONNAISE

Place the red pepper under the broiler until it begins to turn black. You may have to turn it several times to blacken each surface.

When cooled, peel the pepper and remove the seeds. Chop the pepper into pieces.

In the meantime, in a small frying pan, sauté the minced onion in butter until the onion is browned and caramelized.

Place all the pepper, onion, garlic, salt, pepper, and mayonnaise in a food processor or blender. Process or blend the mixture until all the pieces are uniform in size.

Cover and refrigerate until you are ready to use.

the recipes

Part II: The Recipes

CHEVON ENCHILADAS

> You can't even believe how fast these tasty tidbits disappear! I make a double batch of them and never get to taste them myself!

SERVES 12

- 2 POUNDS GROUND CHEVON
- 2 LARGE ONIONS, CHOPPED
- 1½ CUPS COTTAGE CHEESE
- 1½ CUPS SOUR CREAM
- TWO (4-OUNCE) CANS CHOPPED GREEN CHILIES
- ½ TEASPOON CUMIN
- ½ TEASPOON CORIANDER
- TWO (8-OUNCE) CANS TOMATO SAUCE
- 1 CUP SALSA (PAGE 189)
- 1 TABLESPOON CHILI POWDER
- 1 TEASPOON DRIED OREGANO
- ½ TEASPOON GRANULATED GARLIC
- ½ TEASPOON DRIED THYME
- TWELVE (8-INCH) WHOLE WHEAT TORTILLAS
- ¾ CUP SHREDDED CHEDDAR CHEESE

In a large skillet, brown the chevon and half of the onion over medium heat until all the pink is gone. Drain well on paper towel to remove some of the fat. Place in a large bowl and add the cottage cheese, sour cream, chilies, cumin, and coriander, and set aside.

To make the sauce, sauté the remaining onion in a large skillet until it's translucent. Add the tomato sauce, salsa, chili powder, oregano, granulated garlic, and thyme. Simmer uncovered for about 20 minutes or, until the sauce begins to thicken.

Preheat the oven to 350 degrees.

Place approximately ½ cup of the meat mixture on each tortilla, roll up, and place seam side down in two deep 9 x 13 inch baking dishes that are well greased. Pour the sauce over the tops of the enchiladas.

Bake for 30 to 35 minutes. Sprinkle with the cheddar cheese and bake for an additional 5 to 10 minutes, or until the cheese is melted.

CHEVON STROGANOFF

> This is another casserole always served at Goat School, and it's always a huge hit! I double this recipe for Goat School and use a deep 9 x 13 inch pan, and bake it for 1 hour.

SERVES 4 TO 6

ONE (8-OUNCE) PACKAGE EGG NOODLES

2 TABLESPOONS BUTTER

ONE SMALL SWEET ONION
sliced

½ CUP GREEN BELL PEPPERS
diced

1 POUND GROUND CHEVON
or loose sweet goat sausage

ONE (6-OUNCE) CAN TOMATO PASTE

½ CUP SOUR CREAM

¼ TEASPOON SALT

1 CUP WHOLE GOAT MILK RICOTTA
at room temperature (page 79)

ONE (14.5-OUNCE) CAN TOMATO SAUCE

1 SMALL CAN SLICED BLACK OLIVES
drained

Cook the noodles according to the package directions, rinse with cold water, and set aside to drain. In a frying pan, melt the butter, then add the onions and peppers and cook for 2 to 3 minutes. Add the ground chevon or sausage. Brown well, then turn onto paper towels to drain away some of the fat.

Preheat the oven to 350 degrees.

In a small bowl, combine the tomato paste, sour cream, and salt, then set aside. Toss the noodles with the ricotta, and add the tomato paste combination along with the drained meat. Mix this together and put into a well-greased, two-quart casserole dish. Pour the tomato sauce over the top and sprinkle with the black olives.

Bake for 35 to 45 minutes.

CHEVRE "STUFFED" CHICKEN

> This faux stuffed chicken is easy to put together and the taste is phenomenal. You can make this ahead, cover tightly with foil wrap, and freeze! Let it thaw before baking.

SERVES 6 TO 8

1½ POUNDS UNCOOKED CHICKEN TENDERS

4 OUNCES PLAIN CHEVRE
(page 81)

4 TABLESPOONS MINCED HERBS
such as parsley, basil, and chives

½ TEASPOON OLIVE OIL

KOSHER SALT AND FRESHLY GROUND BLACK PEPPER
to taste

½ CUP CHICKEN STOCK

1 TEASPOON BROWN SUGAR

1 EGG

1½ TABLESPOONS LEMON JUICE

Preheat the oven to 350 degrees.

Place the chicken tenders on a foil-covered baking sheet and bake for about 20 minutes, or until just barely cooked.

Meanwhile, in a small bowl, mix together the chevre, mixed herbs, olive oil, and salt and pepper. Add a touch more oil if the cheese isn't really creamy and easy to spread.

In a well-greased casserole dish, place a layer of chicken tenders. Spread the chicken tenders with the cheese mixture, then layer the balance of the tenders on top.

In a small pan, mix the stock and brown sugar. Place over medium heat until simmering. In the meantime, whisk the egg until it's light colored and fluffy. Add the lemon juice to the eggs and mix well. Now add a tiny bit of the simmering stock to the egg mixture to warm it up. Add this warmed lemon and egg to the stock and cook about 2 minutes or until slightly thickened.

Pour over the chicken and bake for 35 minutes.

DYNAMITES

> This is a true Maine recipe and is a "sloppy joe" type meal.

SERVES 10 TO 12

1½ POUNDS GROUND CHEVON
browned

2 LARGE CANS DICED TOMATOES

2 SMALL CANS TOMATO PASTE

2 BUNCHES CELERY
chopped

7 GREEN PEPPERS
chopped

3 MEDIUM ONIONS
chopped

½ TEASPOON DRIED RED PEPPER FLAKES

HAMBURGER OR BULKY ROLLS
(page 123)

Over medium heat, sauté the chevon until browned.

Mix the chevon, tomatoes, tomato paste, celery, green peppers, onions, and red pepper flakes together in a large pot and simmer for at least 4 or 5 hours.

GOAT SAUSAGE PASTA

Fast and easy!

SERVES 5

1 POUND PASTA

1 TEASPOON OLIVE OIL

¼ CUP SEASONED BREAD CRUMBS

1 POUND SPICY GOAT SAUSAGE
(either hot or garlic works great)

1 CLOVE GARLIC
minced

TWO (14.5-OUNCE) CANS DICED TOMATOES

ONE (4–TO–6–OUNCE) CAN SLICED BLACK OLIVES

Cook the pasta according to the package directions.

In the meantime, in a small frying pan, add the olive oil and toast the bread crumbs. Remove from the heat and set aside.

Cut the sausage into ½ inch slices and add to a large skillet. Cook until the sausage is done all the way through. Add the garlic and cook about one minute longer. Stir in the tomatoes and olives and cook, stirring constantly, until the tomatoes are hot.

Add the drained pasta to the skillet, toss all the ingredients together and sprinkle with bread crumbs.

GOAT SCHOOL CHILI

If you have attended Goat School and begged me for this recipe to no avail, well, here it is! Super easy and really good!

SERVES 8 TO 10

2 TABLESPOONS BUTTER

1 POUND GROUND CHEVON

THREE (15.5-OUNCE) CANS BEANS
drained

THREE (15.5 OUNCE) CANS DICED TOMATOES

1 LARGE ONION
chopped

1 LARGE GREEN PEPPER
chopped

2 TABLESPOONS CHILI POWDER

1 TEASPOON CRUSHED RED PEPPER

1 TEASPOON SALT

1 TEASPOON FRESHLY GROUND BLACK PEPPER

In a large pot, melt the butter and add the chevon. Cook thoroughly.

Add the beans and the tomatoes to the cooked chevon and mix. Add the onion, green pepper, chili powder, crushed red pepper, salt, and pepper. Stir well.

Place over medium heat until mixture begins to boil, then turn down to low and cook for 2 or more hours.

Serve with finely chopped onions, shredded cheddar cheese, or just with bread or crackers!

GOAT SCHOOL MAC & CHEESE

> Another Goat School favorite! We always strive to have a meatless dish at each buffet lunch, and this one is always a crowd pleaser. It's creamy, rich, and a true comfort food.

SERVES 8 TO 10

2 CUPS ELBOW MACARONI

2 TABLESPOONS BUTTER

2 TABLESPOONS FLOUR

1 CUP GOAT MILK

¼ TEASPOON SALT

¼ TEASPOON FRESHLY GROUND BLACK PEPPER

3 CUPS SHREDDED CHEDDAR CHEESE

SEASONED BREAD CRUMBS
or croutons

Cook the macaroni according to the package directions for al dente.

Preheat the oven to 350 degrees.

In a large heavy-bottom pot, melt the butter. Add the flour and stir until it's really thick, like paste. Add the milk, salt, and pepper, and continue to stir until you have a nice creamy sauce. Add the cheese and stir until all the cheese is melted. Add the macaroni.

Spray a casserole dish with non-stick spray. Add the mac & cheese. Sprinkle the seasoned bread crumbs over the top, or crush a few croutons and sprinkle them.

Bake for about 30 minutes, or until bubbly.

HONEY-RUBBED CHEVON STEAK

> Chevon is closer in flavor to beef than any other meat. When using in a recipe, however, chevon is cooked very similar to venison because of its very low fat content. Just as an aside, chevon has less cholesterol than turkey or chicken.

SERVES 4 TO 6

½ CUP HONEY

½ CUP LIGHT OLIVE OIL

1 TABLESPOON MINCED GARLIC

1 TABLESPOON CHILI POWDER

1 TABLESPOON SALT

½ TABLESPOON FRESHLY GROUND BLACK PEPPER

2 POUNDS CHEVON STEAK
(the thinner the better)

In a small bowl, mix the honey, olive oil, garlic, chili powder, salt, and pepper. Mix together well, actually mashing the solid ingredients against the sides of the bowl.

Rub the steaks liberally with the marinade. Place in a covered glass dish, pour whatever is left of the marinade over the steaks, and refrigerate overnight.

When ready to cook, allow the steaks to come to room temperature. This will achieve the most even cooking and optimum flavor.

Grill as you would a beef steak to rare, medium rare, or well done.

EASY HOMESTYLE CHEVON AND PASTA

A sure-to-please pasta dish.

SERVES 8 TO 10

1 POUND SWEET OR HOT GOAT SAUSAGE
(or be adventurous and use both!)

1 TABLESPOON OLIVE OIL

1 MEDIUM ONION
finely chopped

1 CLOVE GARLIC
chopped

1 POUND GROUND CHEVON

TWO (32-OUNCE) CANS TOMATO SAUCE

ONE (6-OUNCE) CAN TOMATO PASTE

1 MEDIUM GREEN BELL PEPPER
chopped

1 POUND SPAGHETTI

Preheat the oven to 350 degrees.

In a foil-lined pan, bake the sausage for 35 minutes. When cooled, cut sausage into bite-size pieces.

Put the olive oil in a good-size skillet and gently sauté the onion and garlic until translucent. Add the ground chevon and cook through.

Add the sauce, paste, green pepper, and sausage, stir, then simmer gently for about 2 hours.

Cook the pasta according to the package directions.

Serve the sauce over the cooked spaghetti.

RED BEANS AND RICE

Spicy, full of protein — a great, easy dish.

SERVES 6 TO 8

1 POUND GARLIC GOAT SAUSAGE
cooked and cut into pieces

ONE (30-OUNCE) CAN AND ONE (16-OUNCE) CAN RED BEANS
drained

1 LARGE ONION
chopped

2 CELERY STALKS
chopped

1 LARGE GREEN BELL PEPPER
chopped

1 TEASPOON CAYENNE PEPPER

2 TABLESPOONS CHOPPED PARSLEY

2 TEASPOONS CHOPPED THYME

3 BAY LEAVES

2 CUPS COOKED WHITE RICE

Preheat the oven to 350 degrees.

Bake the sausage in a foil-lined pan at for 35 minutes. When cooled a bit, cut the sausage into bite-size pieces.

Put the beans, onion, celery, green pepper, cayenne pepper, parsley, thyme, bay leaves, and 3 cups of water in a 2-quart pot and simmer for 3 hours.

Remove approximately 1 cup of the cooked mixture and mash with a fork; then add the mash back into the pot along with the pre-cooked sausage and cook for another 30 minutes.

Serve over cooked white rice and garnish with chopped green onion or chives.

SAUSAGE AND VEGETABLE PENNE

> A great, fast summertime dish.

SERVES 2

½ POUND GARLIC GOAT SAUSAGE

2 CUPS PENNE PASTA

2 TABLESPOONS OLIVE OIL

1 MEDIUM-SIZE SWEET ONION
chopped

½ GREEN BELL PEPPER
chopped

1 LARGE CARROT
sliced into long, thin, 2-inch pieces

1 SMALL ZUCCHINI
chopped

1 SMALL SUMMER SQUASH
chopped

1 TEASPOON CHOPPED PARSLEY

1 TEASPOON CRUSHED RED PEPPER

2 TABLESPOONS FINELY CHOPPED CHIVES

Preheat the oven to 350 degrees.

Bake the sausage on a foil-lined pan for 35 minutes. Once cooked and cooled a bit, cut the sausage into bite-size pieces.

Meanwhile, cook the pasta according to the package directions.

Place the olive oil in a large skillet and add the onion, green pepper, and carrot. Stir-fry over medium-high heat for about 3 minutes. Add the zucchini and summer squash and lightly sauté. Add the cooked sausage, parsley, and crushed red pepper, and then stir in the cooked pasta. Stir until everything is hot.

Garnish with chives.

SAUTÉED SAUSAGE OVER ANGEL HAIR

> Who doesn't love a quick stir fry?

SERVES 4 OR 5

½ POUND GOAT SAUSAGE
1 POUND ANGEL HAIR PASTA
1 TABLESPOON OLIVE OIL
1 CUP SLICED MUSHROOMS
3 ROMA TOMATOES
roughly chopped
1 TEASPOON MINCED GARLIC
½ POUND FRESH SPINACH

Preheat the oven to 350 degrees.

Bake the sausage on a foil-lined pan for about 30 minutes. When slightly cooled, cut the sausage into 3-inch pieces.

In the meantime, cook the pasta according to the package directions.

Heat the olive oil in a large frying pan. Add the sausage and stir-fry for about 3 minutes. Then add the mushrooms and tomatoes. When heated through, add the spinach and cook until the spinach is wilted.

Serve over a bed of pasta.

Part II: The Recipes

SPICY CHORIZO CASSEROLE

> This casserole is always a huge hit, both at Goat School and anywhere else I bring it!!

SERVES 4 TO 6

1 POUND PASTA
(I use medium shells or bow ties)

1 TABLESPOON OLIVE OIL

1 POUND LOOSE CHORIZO GOAT SAUSAGE
browned

1 MEDIUM SWEET RED PEPPER
chopped

8 GREEN ONIONS
(green and chopped white parts, chopped)

¼ CUP FLOUR

¼ TEASPOON CHILI POWDER

1 TEASPOON CHIPOTLE PEPPERS IN ADOBO SAUCE
minced

¼ TEASPOON SALT

½ TEASPOON GROUND CUMIN

2¼ CUPS GOAT MILK

1 CUP SHREDDED GOAT MOZZARELLA
(page 76)

2 TABLESPOONS MINCED FRESH CILANTRO

Preheat the oven to 400 degrees.

Cook the pasta according to the package directions.

In a large skillet, add the olive oil and chorizo, and brown over medium heat. Add the pepper and green onions and sauté until they are tender. Stir in the flour, chili powder, chipotle pepper, salt, and cumin, and blend well. Slowly stir in the milk and simmer until it begins to thicken, about 2 minutes. Stir in ¼ of the cheese until the cheese starts to melt.

In a large, well-greased casserole, add the pasta and then stir in the sauce, the sausage, and the cilantro. Sprinkle with the remaining cheese.

Bake uncovered for about 25 minutes, or until it starts to bubble.

STONY KNOLLS SAUSAGE PIZZA CASSEROLE

A nice party alternative to ordering out!

SERVES 8 TO 10

2 POUNDS ZITI, MACARONI
or similar shaped pasta

2 TABLESPOONS BUTTER

2 POUNDS LOOSE GARLIC GOAT SAUSAGE

1 GREEN BELL PEPPER
finely chopped

2 LARGE ONIONS
chopped

½ POUND MUSHROOMS
sliced

THREE (16-OUNCE) JARS PIZZA SAUCE

TWO (6-OUNCE) CANS TOMATO PASTE

1 POUND SHREDDED GOAT MOZZARELLA
(page 76)

½ POUND SLICED PEPPERONI

1 SMALL CAN SLICED BLACK OLIVES
drained

Preheat the oven to 350 degrees.

Cook the pasta according to the package directions for al dente.

Melt the butter in a large skillet and brown the sausage with the green pepper and onions; stir in the mushrooms, pizza sauce, and tomato paste.

Stir the pasta into this mixture and spread into a well-greased, deep 9 x 13 inch pan.

Top with the mozzarella, pepperoni, and black olives.

Bake for 40 minutes.

the recipes

VEGETARIAN GOAT CHEESE LASAGNA

> You will knock the socks off your guests with this dish.

SERVES 12

1 BOX LASAGNA NOODLES

2 POUNDS FRESH SPINACH
stems removed

⅓ CUP OLIVE OIL

1 SMALL SHALLOT
minced

2 LARGE CLOVES GARLIC
very finely minced

3 POUNDS ASSORTED MUSHROOMS

½ CUP DRY WHITE WINE
(you can also use white grape juice)

2 TABLESPOONS CHOPPED FRESH BASIL

1 TABLESPOON CHOPPED FRESH THYME

1 CUP GOAT RICOTTA
(page 79)

KOSHER SALT AND FRESHLY GROUND BLACK PEPPER

½ POUND SHREDDED GOAT MOZZARELLA
(page 76)

10 OUNCES SOFT FRESH CHEVRE
(page 81)

½ TEASPOON NUTMEG

Cook the pasta according to the package directions.

Spray a 9 x 13 inch baking dish with cooking spray. Preheat the oven to 350 degrees.

Soak the spinach in a large bowl of cold water while you start cooking the rest of the ingredients.

Place the oil in a large skillet and add the shallot and garlic and sauté until transparent but not brown. Add the mushrooms and toss gently until coated with the olive oil, making sure you keep stirring very gently until they are reduced by half. This will take 4 to 5 minutes.

Add the wine and simmer, scraping often to make sure all the little brown bits that stick to the pan are included. When most of the liquid has evaporated, remove the pan from the heat and add the basil, thyme, and ricotta. Season to taste with the salt and pepper. Stir in half of the mozzarella. Set this mixture aside.

Shake the water from the spinach and place it in a large sauce pan with only the water that is left on the leaves. Cover and cook at medium heat until it is well wilted, about 5 minutes. Drain the spinach and press it really lightly with the back of a spoon to remove some more of the moisture. Turn this out on a cutting surface and coarsely chop.

In a large bowl, combine the spinach with the remaining mozzarella and half of the chevre. Add the nutmeg and another dash of salt and pepper. Mix until creamy.

Spread one third of the mushroom mixture over the bottom of the baking dish. Top with a layer of noodles, then spread half of the spinach mixture over the noodles. Add another layer of noodles and repeat, ending with a layer of mushrooms. Scatter the remaining chevre on the top and cover the pan tightly with aluminum foil.

Bake covered with foil for 35 to 40 minutes. Remove the foil and place the pan under the broiler until the cheese is browned and bubbling. Let the lasagna rest for a few minutes before serving.

Breads

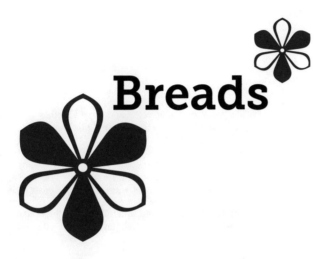

ANADAMA BREAD

> Supposedly the name comes from an old Colonial tale about a man with a lazy wife. One day he got really angry, threw some flour, yeast, cornmeal, and molasses into a bowl and baked his own bread, mumbling, "Anna, damn her."

MAKES 2 LOAVES

1 CUP WATER
at 110 degrees

2 TABLESPOONS YEAST

1 TABLESPOON SUGAR

1 CUP COLD WATER

½ CUP CORN MEAL, PLUS 1 TEASPOON

4 TABLESPOONS BUTTER

2 TEASPOONS SALT

1 CUP MOLASSES

6 CUPS FLOUR

1 EGG
beaten with 2 tablespoons water

Mix the warm water, yeast, and sugar together in a large bowl. Cover and let rest for about 10 minutes.

In the mean time, place the cold water and ½ cup of the corn meal into a small pan over medium heat, stirring continuously until the mixture becomes thick. Remove from the heat and add the butter, salt, and molasses. Mix and let set until the yeast mixture is ready.

Once the yeast is foamy and bubbly, add the flour and the corn meal mixture. Stir well, then turn it out on a floured board and knead for about 10 minutes.

When the dough is nice and smooth, place it in a greased bowl, cover it, and let it sit for about one hour.

When the dough has risen nicely, divide it in half, shape each half into a loaf, and place into 5 x 7 inch greased bread pans. Cover and let the dough rise again for another hour.

Preheat the oven to 350 degrees.

When the bread is ready to go into the oven, paint the tops of the loaves with the egg mixture. Once the loaves are painted, sprinkle the tops with a teaspoon or so of cornmeal.

Bake for 30 minutes.

BURGER BUNS

> Want to really wow your guests at your next cookout? Try these. They are easy to make and taste wonderful!

MAKES 16 TO 18 BUNS

1½ CUPS GOAT MILK

1 STICK BUTTER
at room temperature

2 EGGS
beaten

½ CUP SUGAR

1 TEASPOON SALT

4½ CUPS FLOUR

1½ TABLESPOONS YEAST

Heat the milk for 2 minutes in the microwave.

Mix the milk, butter, eggs, sugar, salt, flour, and yeast together well, then knead for about 10 minutes. Place in a large bowl and cover lightly. Let the dough rise for at least 1 hour.

Punch the dough down and divide it into 16 to 18 segments. Shape each segment into a nice round ball and place on a greased cookie sheet about 2 inches apart. Cover again and let rise about 30 minutes.

Preheat the oven to 350 degrees.

Just before the rolls go into the oven, brush the tops with a little bit of milk. At this point you can sprinkle a few sesame seeds, poppy seeds, or dried onion flakes on them, or just leave them plain.

Bake for 18 to 20 minutes. If you had to use 2 baking sheets, you might want to rotate the sheets after the first 10 minutes.

BUTTERMILK AND DILL BREAD

> This bread makes the most amazing sandwiches! Just lightly toast it and it's great with tuna, any sandwich meat, egg, or anything else!

MAKES 2 LOAVES

½ CUP WATER
at 110 degrees

2 TEASPOONS SUGAR

2 TABLESPOONS YEAST

2 CUPS GOAT BUTTERMILK
(page 77)

¼ CUP OLIVE OIL

¼ CUP DRIED ONION
(you can also use fresh diced onion)

2 TEASPOONS SALT

2 EGGS
lightly beaten

3 TABLESPOONS DILL WEED
(I use dried, but fresh will work also.)

6 CUPS FLOUR

1 EGG WHITE
beaten with 1 tablespoon water

½ TEASPOON DILL SEEDS
optional

Mix the water, sugar, and yeast together in a large bowl. Cover and let set for about 10 minutes or until foamy and bubbly.

When the yeast mixture is ready, add the buttermilk, olive oil, onion, salt, eggs, dill weed, and flour to the bowl and mix. When mixed, turn out onto a floured surface and knead. You may have to add another 1 to 1 ½ cups of flour at this point. Make sure the dough is soft, but not sticky.

Place the dough in a covered bowl, and let it rise about 45 minutes. This dough rises quickly!

Turn the dough out on a floured surface and divide in half. Place each half in a 5 x 7 inch greased loaf pan. Cover again and let rise for another 30 to 45 minutes.

Preheat the oven to 350 degrees.

Paint the loaves with the egg white mixture. Then sprinkle the tops of the loaves with the dill seeds.

Slash the tops of the loaves lengthwise and bake for 35 minutes.

This bread makes the most amazing sandwiches! Just lightly toast it and it's great with tuna, any sandwich meat, egg, or anything else!

BUTTERMILK BREAD

This buttermilk bread has the consistency of a sourdough bread. It makes wonderful toasts and great sandwiches, plus it's really easy to make!

MAKES 2 LOAVES

9½ CUPS FLOUR

4½ TABLESPOONS SUGAR

3 TABLESPOONS YEAST

2½ TEASPOONS SALT

1½ CUPS GOAT BUTTERMILK
(page 77)

⅔ CUP BUTTER

1 EGG
beaten with 2 tablespoons of water

In a large bowl, mix the flour, sugar, yeast, and salt together.

Place the buttermilk, 1½ cups of water, and butter in a bowl and microwave for 2 minutes.

Mix the buttermilk mixture into the flour mixture and stir well. Turn out onto a floured surface and knead for about 10 minutes. Place the dough in a large greased bowl and cover lightly. Let the dough rise for 1 to 1½ hours.

Punch down the dough and divide it in half. Form into loaves and place the loaves in greased 5 x 7 inch bread pans. Cover and let the loaves rise again for another hour.

Preheat the oven to 350 degrees.

Paint the tops of the loaves with the egg mixture. Then score the loaves lengthwise with a sharp knife.

Bake for 30 minutes.

the recipes

Part II: The Recipes

CHEDDAR CHEESE BREAD

> This bread is awesome! It makes great toast, great sandwiches, and if you brush it lightly with olive oil, it grills beautifully.

MAKES 2 ROUND LOAVES

1 CUP WATER
at 110 degrees

1 TABLESPOON YEAST

2 TABLESPOONS SUGAR

1 CUP GOAT MILK

1 TABLESPOON BUTTER

1 TABLESPOON SALT

5 TO 5½ CUPS WHITE BREAD FLOUR
(I recommend using bread flour for this bread rather than all-purpose flour because it will help the bread keep its shape better.)

2 CUPS GRATED CHEDDAR CHEESE

1 EGG
beaten with 1 tablespoon water

Combine the water, yeast, and the sugar in a large bowl. Mix well and cover. Let stand about 5 minutes, or until it's foamy.

In the meantime, heat the milk and butter until warm and the butter is softened. I use the microwave for 1 minute 30 seconds to perform this step.

To the yeast mixture, add the salt and the milk and butter mixture and mix in the flour two cups at a time. When this is barely combined, add the cheese and keep mixing. When it becomes too hard to mix by hand, turn the dough out on a floured surface and begin to knead. Knead about 10 minutes and place in a greased bowl and cover. Set in a warm place to rise for about 1 hour.

After the hour of rising, punch down the dough and divide in half. Shape into two round balls, place on a greased baking sheet, and let rise another hour.

Preheat the oven to 350 degrees. Brush the loaf tops with the egg mixture.

Place the loaves in the oven and bake for 30 to 35 minutes or until a beautiful golden brown.

CINNAMON (RAISIN) BREAD

This tends to be everyone's favorite bread, toasted! The taste and warmth of cinnamon is definitely a comfort food! I can't even begin to tell you how many loaves of this delightful bread I have made and sold throughout the years.

MAKES 3 LOAVES

⅓ CUP WATER
at 110 degrees

1½ TABLESPOONS YEAST

⅓ CUP SUGAR

1½ CUPS GOAT MILK

6 TABLESPOONS BUTTER

7½ CUPS FLOUR

3 EGGS
lightly beaten

1½ TEASPOONS SALT

1 CUP RAISINS
optional

3 EGGS
beaten with 3 tablespoons of water

2 TABLESPOONS CINNAMON

RAW SUGAR
optional (I use raw sugar because it keeps its nice granulated shape throughout the baking).

Place the water, yeast, and ⅓ cup sugar in a large bowl, mix and cover. Let set for about 10 minutes.

Meanwhile, heat the milk and the butter in the microwave for 2 minutes.

When the yeast is all foamy and bubbly, add the eggs, salt, milk and butter, flour, and raisins.

Mix into a dough and then turn the dough out onto a floured surface and knead for about 10 minutes. Place the kneaded dough into a large greased bowl, cover lightly, and set aside to rise for 1 to 1 ½ hours.

When the dough has risen, punch it down and divide it into three equal parts.

On a floured surface, roll the dough out into a large rectangle. Paint the dough with the egg, then sprinkle it with the remaining sugar and cinnamon. Roll the dough up like a jelly roll and place in 5 x 7 inch greased bread pans. Cover the dough and let it rise for another hour.

Preheat the oven to 350 degrees. When the breads are ready to be baked, paint the tops with the leftover egg and sprinkle with raw sugar.

Bake for 25 minutes.

Part II: The Recipes

GARLIC FRENCH BREAD

> This is an amazing, garlicky bread!

MAKES 2 LOAVES

1½ TABLESPOONS YEAST

2 TABLESPOONS SUGAR

2 CUPS WATER
at 110 degrees

1 TABLESPOON SALT

5 TO 5½ CUPS FLOUR

2 TABLESPOONS MINCED GARLIC

2 TABLESPOONS CHOPPED DRIED OREGANO

3 TABLESPOONS YELLOW CORNMEAL

1 EGG
beaten with 2 tablespoons of water

Combine the yeast, sugar, and water in a large bowl. Cover for about 10 minutes. When the yeast mixture is all foamy and bubbly, add the salt, flour, garlic, and oregano.

Mix these ingredients well and knead on a floured surface. Then place the dough in a greased bowl and cover. Let rise for about 1 hour.

Divide the dough into two parts and shape them into long loaves. Place the loaves on a cookie sheet that has been sprinkled with the cornmeal. Paint the tops of the two loaves with the egg mixture.

Take a sharp knife and score the loaf in several places.

Place the loaves in a cold oven. Set the heat to 400 degrees and bake for 35 minutes or until well browned.

HONEY AND WHOLE WHEAT BREAD

A wonderfully hearty bread!

MAKES 3 LOAVES

3 TABLESPOONS YEAST

3 TABLESPOONS SUGAR

3¾ CUPS WATER
at 110 degrees

1½ TABLESPOONS SALT

⅓ CUP VEGETABLE OIL
(I use canola oil.)

¾ CUP HONEY

¾ CUP DRY MILK
(I use store-bought.)

7½ CUPS WHOLE-WHEAT FLOUR

3¾ CUPS WHITE FLOUR

1 EGG
beaten with 2 tablespoons water

Combine the yeast, sugar, and water in a large bowl. Cover and let set about 10 minutes. When the mixture is bubbly and foamy, add the salt, oil, honey, dry milk, and both flours.

Mix well, turn out onto a floured surface, knead for about 10 minutes or until the dough is smooth and shiny. Place in a greased bowl and let rise for at least 1½ hours. Punch down and then divide the dough into three loaves.

Place each loaf in a greased 5 x 7 inch pan, cover it again, and let it rise for another hour.

Preheat the oven to 350 degrees. Brush the loaf tops with the egg mixture. Bake for 30 to 35 minutes or until well browned and hollow sounding when you tap it.

the recipes

SALLY LUNN BREAD

> This is a wonderful, rich white bread that makes great toast and slices nicely for sandwiches! And who was Sally Lunn? Well, she was the daughter of an English pastry chef in Bath, England, in the late 1600s.

MAKES 2 LOAVES

1 TABLESPOON YEAST

¼ CUP WATER
at 110 degrees

2 TABLESPOONS SUGAR

1¾ CUPS GOAT MILK

¼ CUP BUTTER

1 TEASPOON SALT

2 EGGS
well beaten

5 CUPS FLOUR

Place the yeast, water, and sugar in a large bowl, mix it all together, and then cover. Let it rest for about 10 minutes, or until it is foamy and bubbly.

In the meantime, heat the milk and butter to 110 degrees. (I put it in the microwave for about 2 minutes.)

To the yeast mixture, add the salt, the eggs, the milk and butter, and the flour. Mix well and then turn out onto a floured surface and knead for about 10 minutes. After kneading, place it in a greased bowl and cover lightly. Let the dough rise for 1 to 1½ hours.

When the dough has risen, turn it out and divide it in half. Place each half in a greased 5 x 7 inch pan, then cover again and let rise for about 1 hour.

Preheat the oven to 400 degrees.

Bake the loaves for 15 minutes, then reduce the heat to 350 degrees. Bake for an additional 15 to 18 minutes.

SCOTCH OATMEAL BREAD

Because the dough is soft, it's a little difficult to work with, but well worth the trouble!

MAKES 4 LOAVES

5 CUPS WATER
at 110 degrees

3 TABLESPOONS YEAST

1 CUP MOLASSES

4 TABLESPOONS CANOLA OIL

2 TEASPOONS SALT

3 CUPS REGULAR OATMEAL

10 TO 11 CUPS FLOUR

Stir the yeast into the water and add the molasses. Stir well and cover lightly.

Once the yeast mixture is foamy and bubbly, about 10 minutes, add the oil, salt, oats, and flour. Stir into a soft dough. Place in a really big greased bowl, cover lightly, and let rise for about 45 minutes.

Turn the dough out onto a floured surface and divide into four loaves. Place each loaf into a 5 x 7 inch greased bread pan.

Cover the pans and let the loaves rise again for 45 minutes.

Preheat the oven to 350 degrees.

Score the loaves lengthwise with a sharp knife. I usually sprinkle a few oats on the tops of the loaves.

Bake the loaves for 30 minutes.

SEED AND GRAIN BREAD

This bread is a little complicated, but once you make it, you will continue to make batch after batch. It's great toasted and makes wonderful sandwiches, too!

MAKES 2 LOAVES

3 TABLESPOONS SUNFLOWER SEEDS

2 TABLESPOONS WHEAT GERM

2 TABLESPOONS SESAME SEEDS

1 TABLESPOON FLAX SEED

2 TABLESPOONS MILLET

1 TABLESPOON QUINOA

2 TABLESPOONS YEAST

1 TABLESPOON SUGAR

½ CUP WATER
at 110 degrees

1 CUP QUICK-COOKING ROLLED OATS

1½ CUPS BOILING WATER

2 TEASPOONS SALT

⅓ CUP SHORTENING

¼ CUP HONEY

¼ CUP MOLASSES

5½ TO 6 CUPS FLOUR

2 EGGS
beaten

1 EGG
beaten with 2 tablespoons water

Mix the sunflower seeds, wheat germ, sesame seeds, flax seed, millet, and quinoa together in a small bowl and set aside.

Place the yeast, sugar, and warm water in a large bowl. Mix, cover, and let sit for about 10 minutes.

While this mixture is sitting, mix the oats into the boiling water. Then add the salt, shortening, honey, and molasses.

When the yeast mixture is foamy and bubbly, add the flour and the seed and grain mixture, and then lastly add the oat mixture.

Stir all of this into a dough and turn out onto a floured surface. Knead for about 10 minutes. Place the dough into a floured bowl, cover it lightly, and let it set for about one hour.

Punch the risen dough down and divide it in half. Shape the dough into 2 loaves and place each loaf into a greased 5 x 7 inch bread pan. Cover and let the dough rise again for about 1 hour.

Preheat the oven to 350 degrees. When the dough has risen, paint the tops with the egg mixture. Take a sharp knife and score the top of each loaf lengthwise.

Bake the loaves for 30 minutes.

BANANA NUT BREAD

This has been my favorite quick bread since I was a child. I can remember eating it too fast and getting wicked cases of hiccups! It's a very solid, heavy type of bread that slices beautifully when cooled. Spread it with a little plain chevre (page 81) for a real treat!

MAKES 2 LOAVES

- 1 CUP SUGAR
- ½ CUP SHORTENING
- 2 EGGS
- 1⅓ CUPS MASHED OVER-RIPE BANANAS
- 2½ CUPS FLOUR
- ½ TEASPOON SALT
- 1 TEASPOON BAKING POWDER
- ⅓ CUP CHOPPED WALNUTS

Preheat the oven to 350 degrees.

Put sugar, shortening, eggs, bananas, flour, salt, baking powder, and walnuts in a large bowl and blend well by hand.

Place into two greased and floured 5 x 7 inch loaf pans.

Bake for 1 hour, or until the tops are light brown and crusty.

BLUEBERRY LEMON BREAD

> If you put out a platter of sliced breads this will be the first to disappear.

MAKES 1 LOAF

⅓ CUP BUTTER
melted

1 CUP SUGAR

3 TABLESPOONS LEMON JUICE

2 EGGS

1½ CUPS FLOUR

1 TEASPOON BAKING POWDER

1 TEASPOON SALT

½ CUP GOAT MILK

2 TABLESPOONS GRATED LEMON ZEST

1 CUP FRESH OR FROZEN BLUEBERRIES

Preheat the oven to 350 degrees.

In a large bowl, beat together the butter, sugar, lemon juice, and eggs. In a separate bowl combine the flour, baking powder, and salt, and stir into the butter/egg mixture alternately with the goat milk.

Add the lemon zest and then gently fold in the blueberries.

Place the batter in a well greased 5 x 7 inch loaf pan.

Bake for 60 to 70 minutes. Cool about 10 minutes before turning out of the pan.

You can also bake this bread in mini loaf pans! Use four well greased 2½ x 5 inch pans and bake for 45 minutes.

CHOCOLATE CHIP PUMPKIN BREAD

This is soooo good!

MAKES 2 LOAVES

⅔ CUP APPLESAUCE

2⅔ CUPS SUGAR

4 EGGS

ONE (15.5-OUNCE) CAN PUMPKIN

3⅓ CUPS FLOUR

2 TEASPOONS BAKING SODA

1½ TEASPOON SALT

½ TEASPOON BAKING POWDER

1 TEASPOON CINNAMON

1 TEASPOON CLOVES

1 CUP CHOCOLATE CHIPS

Preheat the oven to 350 degrees.

Grease two 5 x 7 inch loaf pans.

In a large bowl, blend the applesauce with the sugar. Stir in the eggs one at a time. Add the pumpkin and ⅔ cup of water and mix well. Blend in the flour, baking soda, salt, baking powder, cinnamon, and cloves. Finally, fold in the chocolate chips.

Pour into the pans and bake for 70 minutes. Cool for at least 10 minutes before turning out of the pans.

COCONUT BREAD

This bread is so unusual that people can't wait to try it. It's fairly dense and slices beautifully when completely cool.

MAKES 1 LOAF

3 CUPS FLOUR
3 TEASPOONS BAKING POWDER
½ TEASPOON SALT
1 CUP SUGAR
1 CUP FLAKED COCONUT
1 EGG
1 CUP GOAT MILK
1 TEASPOON VANILLA

Preheat the oven to 350 degrees.

Mix the flour, baking powder, salt, sugar, and coconut together in a large bowl. Add the egg, goat milk, and vanilla, and blend well.

Place in a well-greased 5 x 7 inch loaf pan.

Bake for 45 minutes.

CRANBERRY-ORANGE WALNUT BREAD

> The tangy cranberries along with the orange flavor is very pleasant.

MAKES 1 LOAF

- 2 CUPS FLOUR
- ¾ CUP SUGAR
- 1½ TEASPOONS BAKING POWDER
- ½ TEASPOON BAKING SODA
- ¾ TEASPOON SALT
- ¾ CUP ORANGE JUICE
- ZEST OF 1 ORANGE
- ¾ CUP GOAT BUTTERMILK (page 77)
- 1 LARGE EGG, beaten
- 3 TABLESPOONS VEGETABLE OIL
- 1 CUP ROUGHLY CHOPPED FRESH CRANBERRIES
- ½ CUP ROUGHLY CHOPPED WALNUTS

Preheat the oven to 350 degrees.

In a large mixing bowl, whisk together the flour, sugar, baking powder, baking soda, and salt.

In a medium bowl, combine the orange juice, orange zest, buttermilk, egg, and oil.

Mix together both the wet and dry ingredients and stir until well combined. Stir in the cranberries and the nuts.

Pour into a well-greased 5 x 7 inch loaf pan and bake for 55 to 65 minutes. Let the bread cool 10 to 15 minutes before turning out.

DATE NUT BREAD

> Date nut bread is one of my favorites! Spread it with a little plain chevre and it's delicious.

MAKES 1 LOAF

1 CUP CHOPPED DATES

1 TEASPOON BAKING SODA

¾ CUP BOILING WATER

¾ CUP SUGAR

¼ TEASPOON SALT

1 TABLESPOON BUTTER
melted

1¾ CUPS FLOUR

1 EGG

½ CUP CHOPPED WALNUTS

Preheat the oven to 350 degrees.

Put the dates in a medium bowl and sprinkle with the baking soda. Pour the boiling water over them and let cool.

When cooled, add the sugar, salt, melted butter, flour, and egg. Mix well and then fold in the nuts.

Place in a greased and floured 5 x 7 inch loaf pan and bake for 40 minutes. Cool 10 minutes or so before turning out.

PEANUT BUTTER BANANA BREAD

> Maybe we should call this Elvis Bread? Peanut butter and banana is such a great combination!

MAKES 1 LOAF

1¾ CUPS FLOUR
¼ TEASPOON BAKING SODA
2 TEASPOONS BAKING POWDER
½ TEASPOON SALT
⅓ CUP SHORTENING
⅔ CUP SUGAR
¾ CUP PEANUT BUTTER
2 EGGS
beaten
1 CUP MASHED OVER-RIPE BANANAS

Preheat the oven to 350 degrees.

Blend together the flour, baking soda, baking powder, and salt, and set aside. In another bowl, cream the shortening, sugar, peanut butter, and eggs together.

Add the dry ingredients alternately with the bananas to the creamed shortening and peanut butter mixture.

Pour into a well-greased and floured 5 x 7 inch loaf pan.

Bake for 1 hour. Let cool approximately 10 minutes before turning out of the pan.

the recipes

ZUCCHINI BREAD

> Don't you just love recipes that use these excessive green monsters from your garden? There are people who make sure to lock their vehicles at church in fear of coming out and finding boxes full of this over-producing, garden-hogging squash!

MAKES 2 LOAVES

3 EGGS

1 CUP VEGETABLE OIL

2 CUPS SUGAR

2 CUPS GRATED ZUCCHINI
(use the peel also, but discard the seeds of really big zucchinis)

3 CUPS FLOUR

1 TEASPOON BAKING SODA

1 TEASPOON SALT

3 TEASPOONS CINNAMON

½ CUP GOAT MILK

¼ CUP CHOPPED WALNUTS
optional

Preheat the oven to 350 degrees.

In a large bowl, mix the eggs, oil, sugar, and zucchini together. In a separate bowl, blend the flour, baking soda, salt, and cinnamon together.

Add the dry ingredients alternately with the goat milk to the egg and sugar mixture.

Fold in the nuts if desired.

Place in two well-greased and floured 5 x 7 inch loaf pans.

Bake for 1 hour. Let cool 10 to 15 minutes before turning out.

Note: You can also use approximately 5 or 6 mini pans and bake them for 45 minutes.

ONION FLAX CRACKERS

> If you have homemade cheese why not make homemade crackers to go with it?

MAKES 2 LOAVES

2 CUPS FLOUR

1 TABLESPOON SUGAR

½ TEASPOON SALT

2 TEASPOONS ONION POWDER

1 TEASPOON GARLIC POWDER

¼ TEASPOON CAYENNE PEPPER

2 TABLESPOONS BUTTER

¾ CUP FLAX SEED

⅔ CUP WATER

ADDITIONAL FLOUR
for dusting

½ TABLESPOON KOSHER SALT

Preheat the oven to 400 degrees.

Place the flour, sugar, salt, onion powder, garlic powder, and cayenne pepper in a large bowl and mix well. Cut in the butter. Add the flax seed and mix well. Add the water, stirring until the flour is completely moist. Divide the dough into two parts.

Place each half on a lightly oiled 12 x 14 inch sheet of aluminum foil. Pat the dough into 8-inch rounds.

Dust a rolling pin generously with flour and roll the dough to about 1/16 inch thick. Cut out a rectangle about 8 x 12 inches and score it with a knife or pizza cutter into 2-inch squares. Do not cut through the dough.

Place the foil and dough on a cookie sheet and sprinkle with half of the kosher salt. Repeat with the remainder of the dough.

Bake for 7 to 9 minutes or until the edges of the crackers are light golden.

Lift the foil carefully and turn it over so that the dough is face down on the cookie sheet. Peel off the foil and bake the dough an additional 7 to 9 minutes or until the crackers are beige to a very light brown.

Remove them from the oven and place on a rack to cool. Break into squares when completely cool.

Store in a tightly closed container.

Breakfast

CHRISTMAS MORNING STICKY BUNS

> These make a fantastic gift for others or yourself! And you don't have to wait until Christmas to make them!

MAKES 24 ROLLS

FOR THE ROLLS:

¾ CUP SUGAR

2 TABLESPOONS YEAST

½ CUP WATER
at 110 degrees

½ CUP GOAT MILK
at 110 degrees

1 TEASPOON SALT

2 EGGS
beaten

5 CUPS FLOUR

½ CUP BUTTER
melted

2 TEASPOONS CINNAMON

FOR THE TOPPING:

4 TABLESPOONS BUTTER

½ CUP BROWN SUGAR

2 TABLESPOONS KARO SYRUP

½ CUP PECAN HALVES
optional

Measure out ½ cup of the sugar, then remove 1 tablespoon of the sugar and add the yeast and water in a large bowl and let it set for about 10 minutes or until foamy and bubbly. Add the goat milk, the rest of the ½ cup of sugar, salt, eggs, and flour and mix well. Turn out onto a floured surface and knead for about 10 minutes. Place in a large well-greased bowl and cover. Let the dough rise for 1 to 1½ hours.

On a floured surface, turn out the dough and roll it out into a rectangle.

Paint the rectangle with the melted butter, then sprinkle it well with ¼ cup of sugar mixed with cinnamon. Roll the dough up like a jelly roll and slice into 1 to 1½ inch slices. You will get 24 slices out of this roll.

Now make the topping. Combine 2 tablespoons of butter, ¼ cup brown sugar, and 1 tablespoon Karo syrup in each of the two 9 inch cake pans. Place the pans on your stove burners set on the lowest setting and stir until the butter is melted.

Mix this well and make sure it's spread all over the bottom of the pan. Remove from heat.

If you want to put pecans on your sticky buns, now is the time to put some on top of the syrup mixture.

Now count how many little rolls you have, divide in half, and carefully place half into each cake pan.

At this point, you can let them rise for an hour. Then preheat the oven to 375 degrees and bake for 25 to 30 minutes.

Alternatively, you can place the pan in the center of a sheet of plastic wrap, tie up with a ribbon, and freeze! Make sure you put these instructions with the buns for eating on Christmas morning: Remove from freezer the night before Christmas. Remove the plastic wrap and cover lightly with a cloth. Leave on the countertop overnight. In the morning, preheat the oven to 375 degrees and bake for 25 to 30 minutes!

SAUSAGE BREAKFAST CASSEROLE

This casserole needs to be prepared the night before. I usually make it with Honey and Whole-Wheat Bread (page 129), and it's great!

SERVES 6 TO 8

1½ POUNDS LOOSE GOAT BREAKFAST SAUSAGE, UNCOOKED
(see below)

9 LARGE EGGS

3 CUPS GOAT MILK

½ TEASPOON SALT

½ TEASPOON PEPPER

1½ TEASPOONS DRY MUSTARD

12 SLICES BREAD
cut into cubes

1½ CUPS SHREDDED CHEESE
either queso fresco (page 80) or cheddar work well

PAPRIKA
for sprinkling

Brown the sausage and drain really well on paper towels.

In a large bowl, beat the eggs and then add the milk, salt, pepper, and mustard. Mix well.

In a deep 9 x 13 inch pan that has been sprayed well with cooking spray, place the bread cubes. On top of the bread cubes, spread out the cheese, then the sausage. Pour the milk and egg mixture over the top. Cover the pan and place in the refrigerator overnight.

About ½ hour before serving, preheat the oven to 350 degrees and place the casserole on the counter to warm up a bit. Sprinkle with paprika.

Bake the casserole for 45 minutes. This casserole is best if it's allowed to stand for about 10 minutes before serving.

{CHEVON BREAKFAST SAUSAGE}

This sausage mixture can be used in any recipe that requires loose breakfast sausage. Or you can make this into delicious sausage patties!

1 POUND GROUND CHEVON
(goat meat)

½ TABLESPOON KOSHER SALT

½ TEASPOON CHOPPED SAGE

⅓ TEASPOON CHOPPED SUMMER SAVORY

⅛ TEASPOON GROUND NUTMEG

⅔ TEASPOON GROUND MARJORAM

⅓ TEASPOON FRESHLY GROUND BLACK PEPPER

Mix the chevon, salt, sage, savory, nutmeg, marjoram, and pepper together well. Add to any recipe requiring raw sausage meat and cook according to the recipe, or form the meat into patties and cook in a non-stick frying pan.

SOUR CREAM COFFEE CAKE

This coffee cake is not only delicious, but it's also really classy. The layer of topping in the center really sets it apart!

SERVES 8

1 STICK BUTTER
softened

2 EGGS
beaten

1 TEASPOON VANILLA

2 CUPS FLOUR

¼ TEASPOON SALT

1 TEASPOON BAKING SODA

1¼ CUPS SUGAR

½ PINT SOUR CREAM

1 TEASPOON BAKING POWDER

½ CUP CHOPPED NUTS

1 TEASPOON CINNAMON

Preheat the oven to 350 degrees.

In a large bowl, place the butter, eggs, vanilla, flour, salt, baking soda, 1 cup of the sugar, sour cream, and baking powder. Mix well.

Grease a tube pan, and then place half of the batter in the pan.

Mix together the remaining sugar, chopped nuts, and cinnamon, and divide in half. Place half of the topping on top of the batter. Put the rest of the batter in the pan and sprinkle with the rest of the topping.

Bake for 40 to 45 minutes.

TEX-MEX OMELET

> This is one of my husband's favorite breakfasts, especially if it's made with my homemade salsa (page 189)!

SERVES 1

2 LARGE EGGS

2 TABLESPOONS GOAT MILK

1 OUNCE SHREDDED GOAT MOZZARELLA
(page 76)

1½ TABLESPOONS SALSA

Beat the eggs and milk together, then place in a small non-stick frying pan over medium heat. As it cooks, I lift the edges a bit and let the uncooked egg mixture run under and get cooked.

When the eggs are cooked, place the mozzarella on the eggs and allow a minute or so for it to start to melt.

Place the salsa on one half of the eggs and fold the other half over. It's ready!

VEGGIE AND CHEVRE OMELET

> This is a great omelet! I have used garlic and onion chevre, sundried tomato and basil or horseradish and dill chevre in it (page 81).

SERVES 1

¼ CUP DICED ROMA TOMATO

1 GREEN ONION
green and white parts, well diced

1 SMALL MUSHROOM
chopped

2 EGGS
well beaten

1 TABLESPOON GOAT MILK

2 TABLESPOONS CHEVRE
(garlic and onion is really good!) (page 81)

Place the tomato, green onion, and mushroom in a small bowl and heat in the microwave for 30 seconds.

Mix the eggs with the milk and place in a non-stick frying pan over medium heat. As it cooks, lift the edges a bit and let the uncooked egg mixture run under and get cooked.

When the eggs are cooked, spread the chevre over the top. Place the warmed veggies on one half of the eggs and fold over.

Cookies

CHOCOLATE CHIP COOKIES

> I sell a large version of these at the farmer's market and they disappear rapidly! A regular size ice cream scoop makes a nice large cookie. This bigger version must be cooked for 15 minutes.

MAKES 72 COOKIES

2 STICKS BUTTER
softened

¾ CUP GRANULATED SUGAR

¾ CUP BROWN SUGAR

1 TEASPOON VANILLA

2 EGGS

2½ CUPS FLOUR

1 TEASPOON BAKING SODA

1 TEASPOON SALT

½ CUP COARSELY CHOPPED WALNUTS

1 CUP CHOCOLATE CHIPS

Preheat the oven to 350 degrees.

Cream the butter, sugars, and vanilla together until smooth and creamy. Add the eggs one at a time. Gradually add the flour, baking soda, and salt, and mix until they are incorporated. When smooth, add the nuts and chocolate chips.

Drop the batter by rounded teaspoonfuls onto an ungreased cookie sheet. Bake for 8 to 10 minutes until golden brown.

CHOCOLATE CHIP OATMEAL COOKIES

I don't make these cookies very often, because when I do, I tend to eat nearly the whole batch myself! They are really, really good.

MAKES 72 COOKIES

- 1 CUP SHORTENING
- 1 CUP BROWN SUGAR
- 1 CUP GRANULATED SUGAR
- 2 EGGS
- 1 TEASPOON VANILLA
- 2 CUPS FLOUR
- 2 CUPS ROLLED OATS
- 1 TEASPOON BAKING POWDER
- 1 TEASPOON BAKING SODA
- ½ TEASPOON SALT
- ¼ CUP COARSELY CHOPPED WALNUTS
- 1 CUP CHOCOLATE CHIPS

Preheat the oven to 350 degrees.

Cream the shortening and sugars together until smooth and creamy. Add the eggs one at a time, then add the vanilla and blend well. Add the flour, oats, baking powder, baking soda, and salt, and mix until well blended. Fold in the nuts and chocolate chips.

Drop the batter by tablespoonfuls onto a greased cookie sheet, about 2 inches apart.

Bake for 12 to 15 minutes, or until golden brown.

CHOCOLATE CRINKLES

> If you like chocolate, these cookies are the ones for you. Bet you can't eat just one!

MAKES 72 COOKIES

½ CUP VEGETABLE OIL

4 SQUARES UNSWEETENED BAKING CHOCOLATE
melted

2 CUPS GRANULATED SUGAR

4 EGGS

2 TEASPOONS VANILLA

2 CUPS FLOUR

2 TEASPOONS BAKING POWDER

½ TEASPOON SALT

1 CUP CONFECTIONERS' SUGAR

Mix the oil, chocolate, and sugar together. Blend in one egg at a time until well mixed, and then add the vanilla.

Stir the flour, baking powder, and salt into the chocolate mixture. Cover the bowl with plastic wrap and refrigerate for several hours or overnight.

Preheat the oven to 350 degrees. Drop teaspoonfuls of dough into a bowl containing the confectioners' sugar and then shape into balls. Place about 2 inches apart on a greased cookie sheet.

Bake for 10 to 12 minutes.

JOE FROGGERS (BIG MOLASSES COOKIES)

> These cookies date back to Colonial times!

MAKES 36 LARGE COOKIES

- ½ CUP SHORTENING
- 1 CUP SUGAR
- 1 CUP MOLASSES
- 2½ CUPS FLOUR
- 1½ TEASPOONS GROUND GINGER
- 1 TEASPOON BAKING SODA
- 1 TEASPOON SALT
- ½ TEASPOON GROUND CLOVES
- ½ TEASPOON NUTMEG
- ½ TEASPOON ALLSPICE

Cream the shortening and sugar together, then add the molasses and blend well. Add the flour, ginger, baking soda, salt, cloves, nutmeg, and allspice; mix well.

Divide the dough in half and wrap each half in plastic wrap. Chill for at least 1 hour.

Preheat the oven to 350 degrees.

Roll out one half of the dough at a time to about ¼ inch thick. Cut with a 3-inch round cookie cutter and place on a lightly greased cookie sheet.

Bake 10 minutes. Cool 1 to 2 minutes on the pan before lifting off. Let cool on a wire rack. Store cookies in an airtight container.

In Marblehead, Massachusetts, there was a couple named Aunt Creasie and Uncle Joe. Each May, during the elections, they would invite all the inhabitants of town and serve big, flat, chewy molasses cookies. In May, all of the peepers would come out and make lots of noise, thus the frog part! Eventually all of the fishermen from surrounding towns would pay Uncle Joe to make these for them. They're still popular in New England today, but few people know their real name!

MOLASSES COCONUT CHEWS

> **Warning: These cookies are very highly addictive, especially when warm from the oven!**

MAKES 48 COOKIES

1 CUP SUGAR

1 CUP BROWN SUGAR

1 CUP SHORTENING

2 EGGS
beaten

2 TEASPOONS VANILLA

¼ CUP MOLASSES

4 CUPS FLOUR

1 TEASPOON SALT

1½ TEASPOONS BAKING SODA

1 CUP SHREDDED COCONUT

Preheat the oven to 375 degrees.

Cream together both sugars and the shortening. Add the eggs, vanilla, and molasses, and stir well.

Sift together the flour, baking soda, and salt, then stir that into the eggs and sugar. Add the coconut and mix really well.

Shape into balls about the size of a walnut and place on a greased cookie sheet about 3 inches apart. Bake for 12 to 15 minutes.

MOLASSES GINGERSNAPS

> These are really easy cookies to make and go really fast. I tend to be very heavy handed when it comes to adding spices to cookies. I like them very spicy, if you don't, stick to the amounts in the recipe. If you do, double the amount of spices!

MAKES 48 COOKIES

¾ CUP BUTTER
softened

1 CUP SUGAR
plus extra for rolling

¼ CUP MOLASSES

1 EGG

2 CUPS FLOUR

1 TEASPOON BAKING SODA

¼ TEASPOON SALT

1 TEASPOON CINNAMON

1 TEASPOON GROUND CLOVES

1 TEASPOON GROUND GINGER

Preheat the oven to 375 degrees.

Mix the butter, sugar, molasses, egg, flour, baking soda, salt, cinnamon, ground cloves, and ginger together until well blended. Form into walnut-sized-balls and roll each one in sugar.

Place 2 inches apart on a greased cookie sheet. Bake for 8 minutes, or until a nice golden brown.

PEANUT BUTTER COOKIES

> My favorite cookies!

MAKES 48 COOKIES

½ CUP SHORTENING
½ CUP PEANUT BUTTER
1 EGG
½ CUP GRANULATED SUGAR
½ CUP BROWN SUGAR
1¼ CUPS FLOUR
½ TEASPOON BAKING POWDER
¾ TEASPOON BAKING SODA
¼ TEASPOON SALT

Preheat the oven to 375 degrees.

Mix together the shortening, peanut butter, egg, granulated sugar, and brown sugar. Add the flour, baking powder, baking soda, and salt. Blend well.

Roll into balls about the size of a walnut and place 3 inches apart on a greased cookie sheet. Flatten each ball with a fork dipped in sugar, making a criss-cross pattern.

Bake for 10 to 12 minutes

SNICKERDOODLES

> These were the very first cookies I ever made for my husband. He was hooked . . . .

MAKES 72 COOKIES

1 STICK BUTTER
softened

½ CUP SHORTENING

1½ CUPS SUGAR PLUS 2 TABLESPOONS

2 EGGS

2¾ CUPS FLOUR

2 TEASPOONS CREAM OF TARTAR

1 TEASPOON BAKING SODA

¼ TEASPOON SALT

2 TEASPOONS CINNAMON

Preheat the oven to 400 degrees.

Mix together the butter, shortening, 1½ cups of the sugar, and eggs thoroughly.

Sift together the flour, cream of tartar, baking soda, and salt. Stir this into the shortening mixture.

Mix the remaining 2 tablespoons of sugar and cinnamon together in a small bowl. Shape the dough into 1-inch balls and roll in the cinnamon and sugar mixture.

Place on a greased cookig sheet and bake for 8 to 10 minutes.

SOFT AND CHEWY OATMEAL COOKIES

These cookies are a crowd pleaser for sure!

MAKES 48 COOKIES

2 CUPS BROWN SUGAR

1 CUP SHORTENING

2 EGGS
beaten

2 TEASPOONS VANILLA

2 CUPS FLOUR

1½ TEASPOONS BAKING SODA

1 TEASPOON SALT

3 CUPS QUICK OATS

½ CUP GRANULATED SUGAR

Preheat the oven to 375 degrees.

Cream together the brown sugar and shortening. Add the eggs and vanilla and mix really well. Stir in the flour, baking soda, and salt. Mix well. Add the oats and blend.

Take walnut-size pieces of dough and roll into a ball. Roll the balls in the granulated sugar.

Place the balls on a greased cookie sheet about 3 inches apart. Bake for 12 to 15 minutes.

SUGAR COOKIES

> This was my grandmother's recipe. I make them whenever we have groups of children visit the farm. I have a goat-shaped cookie cutter that always makes a big hit!

MAKES 96 COOKIES

3 STICKS BUTTER
softened

2 CUPS SUGAR

4 EGGS

1 TABLESPOON VANILLA

5½ CUPS FLOUR

2 TEASPOONS BAKING POWDER

2 TEASPOONS SALT

2 TEASPOONS NUTMEG OR, BETTER YET, MACE

Cream the butter and sugar together. Add the eggs one at a time, mixing well after each addition, then add the vanilla. Sift together the flour, baking powder, salt, and nutmeg or mace, and add to the egg mixture, and blend well.

Separate into three parts and wrap each part in plastic wrap. Refrigerate for at least 8 hours.

Preheat the oven to 400 degrees.

Taking one part of the dough at a time, place the dough on a floured surface and roll out to ½ inch thick. Cut out desired shapes and place on an ungreased cookie sheet. These cookies spread very little, so they can be placed about ½ an inch apart. Bake for 8 minutes.

Part II: The Recipes

Cakes & Other Treats

BREAD PUDDING

This is true comfort food! Serve with a little scoop of vanilla ice cream—heavenly!

SERVES 4

2 CUPS STALE BREAD
cubed

1 QUART GOAT MILK

2 EGGS

½ CUP SUGAR

1 TEASPOON VANILLA

¼ TEASPOON SALT

½ CUP RAISINS
optional

Preheat the oven to 300 degrees.

Soak the bread in the goat milk until soft and mushy. Mash it a bit, and then heat the milk and bread mixture until nearly boiling in a saucepan over medium-high heat.

Beat the eggs, add the sugar, vanilla, and salt, and add to the bread and milk mixture. Add raisins if desired.

Pour into an 8 x 8 inch baking dish and set the dish in a larger dish of water.

Bake for 45 minutes or until a light golden brown.

BUMPY APPLE CAKE

> This little cake is an autumn favorite. It's soft on the inside and crunchy on the outside. Yum!

SERVES 9

- 1 CUP SUGAR
- ¼ CUP SHORTENING
- 1 EGG
- 1 TEASPOON VANILLA
- ½ TEASPOON BAKING SODA
- 1 CUP FLOUR
- ½ TEASPOON BAKING POWDER
- ½ TEASPOON SALT
- ½ TEASPOON CINNAMON
- ½ TEASPOON NUTMEG
- 3 CUPS PEELED AND DICED APPLES
- ½ CUP CHOPPED WALNUTS

Preheat the oven to 350 degrees.

In a large bowl, mix together the sugar, shortening, egg, vanilla, baking soda, flour, baking powder, salt, cinnamon, and nutmeg. When it is all smooth and creamy, add the apples and chopped walnuts.

Place this mixture in a greased 8 x 8 inch pan and bake for 45 minutes.

CHOCOLATE ZUCCHINI CAKE

A great way to use up some of that prolific green vegetable that seems to appear in your garden overnight!

SERVES 12 TO 15

- 2½ CUPS FLOUR
- ¼ CUP BAKING COCOA
- 1 TEASPOON BAKING SODA
- ½ TEASPOON SALT
- ½ TEASPOON CINNAMON
- ½ TEASPOON GINGER
- ½ CUP VEGETABLE OIL
- 1 STICK BUTTER
  at room temperature
- 1¾ CUPS SUGAR
- 2 EGGS
- 1 TEASPOON VANILLA
- ½ CUP GOAT BUTTERMILK
  (page 77)
- 2 CUPS GRATED ZUCCHINI
  (leave the peel on, however, remove the seeds before grating)
- ¼ CUP CHOPPED WALNUTS
- ¼ CUP CHOCOLATE CHIPS

Preheat the oven to 325 degrees.

Sift together the flour, cocoa, baking soda, salt, cinnamon, and ginger and set aside. In a large bowl, cream together the oil, butter, and sugar. Add the eggs, vanilla, and milk. Mix well, then stir in the dry ingredients. Add the zucchini and blend.

The walnuts and chocolate chips can either be sprinkled on the top of the batter or stirred.

Bake in a well-greased 13 x 9 inch pan for 45 minutes.

DARK CHOCOLATE BUTTERMILK CAKE

Rich, moist, chocolatey goodness — always a hit!

SERVES 12 TO 15

1¾ CUPS FLOUR
2 CUPS SUGAR
¾ CUP BAKING COCOA
1½ TEASPOONS BAKING POWDER
1½ TEASPOONS BAKING SODA
1 TEASPOON SALT
2 EGGS
1 CUP GOAT BUTTERMILK
(page 77)
½ CUP VEGETABLE OIL
2 TEASPOONS VANILLA
1 CUP BOILING WATER

Preheat the oven to 350 degrees. Grease and flour two 9 inch round baking pans.

In a large bowl, stir together the flour, sugar, cocoa, baking powder, baking soda, and salt. Add the eggs, buttermilk, oil, and vanilla.

Mix at medium speed for 2 minutes. Stir in the boiling water. The batter will be very thin. Pour into the prepared baking pans.

Bake for 30 to 35 minutes. Cool for 10 minutes before turning out of the pans onto a wire rack.

Frost with amazing frosting (recipe below).

{AMAZING FROSTING}

½ CUP GOAT BUTTERMILK
(page 77)
2½ TABLESPOONS FLOUR
¼ CUP SHORTENING
¼ CUP BUTTER
at room temperature (do not substitute margarine!)
⅛ TEASPOON SALT
½ CUP SUGAR
½ TEASPOON VANILLA

In a small pan, cook together the buttermilk and flour over low heat until it is a very, very thick paste. Let it cool completely.

In a mixing bowl, cream together the shortening, butter, salt, and sugar. When well blended, add the flour and buttermilk mixture.

Beat at high speed for 5 to 7 minutes, then add the vanilla, and mix another minute.

Part II: The Recipes

GOAT SCHOOL CHERRY CUPCAKES

> These cupcakes are awesome when frosted with chocolate frosting! They really go fast.

MAKES 24 CUPCAKES

1 CUP MARASCHINO CHERRIES
drained, cut into quarters

⅔ CUP RESERVED CHERRY JUICE

2½ CUPS FLOUR

2½ TEASPOONS BAKING POWDER

½ TEASPOON CINNAMON

½ TEASPOON SALT

1⅓ CUP SUGAR

2 STICKS BUTTER
at room temperature

4 EGGS

1 TEASPOON ALMOND EXTRACT

Preheat the oven to 350 degrees and prepare 24 muffin cups with paper liners.

While measuring out the cherries, mix the reserved cherry juice with ⅔ cup of water, and set aside. In a large bowl, combine the flour, baking powder, cinnamon, and salt. In a mixing bowl, beat the sugar with the butter at high speed until light and fluffy, 2 to 3 minutes.

Beat in the eggs and the almond extract until combined. Add the flour mixture alternately with the cherry juice. Fold in the cherries.

Pour the batter evenly into the muffin cups.

Bake for 20 minutes. Let cool in pan about 5 minutes and then turn out onto a rack. Cool completely before frosting.

{CHOCOLATE BUTTERCREAM FROSTING}

6 TABLESPOONS BUTTER
softened

2⅔ CUPS CONFECTIONERS' SUGAR

½ CUP BAKING COCOA

⅓ CUP GOAT MILK

1 TEASPOON VANILLA

In a medium bowl, beat the butter until smooth and light colored. Blend in the confectioners' sugar and cocoa alternately with the milk, beating well after each addition, until the frosting looks nice and spreadable. Add the vanilla and a bit of additional milk if the frosting is a little stiff.

MAPLE CUPCAKES

There is nothing quite like real maple syrup to make mouth-watering goodies.

MAKES 18 CUPCAKES

2½ CUPS FLOUR

2 TEASPOONS BAKING POWDER

1 TEASPOON BAKING SODA

½ TEASPOON SALT

¾ TEASPOON GROUND GINGER

1 STICK UNSALTED BUTTER
softened

½ CUP BROWN SUGAR

2 EGGS

1¼ CUPS MAPLE SYRUP

2 TEASPOONS VANILLA

½ CUP GOAT BUTTERMILK
(page 77)

½ CUP FINELY CHOPPED WALNUTS

Preheat the oven to 350 degrees.

Sift together the flour, baking powder, baking soda, salt, and ginger, and set aside. Cream the butter and brown sugar together until fluffy. Beat in the eggs one at a time, then add the syrup and vanilla.

Stir the flour mixture in alternately with the buttermilk. Fold in the nuts.

Line cupcake pans with papers. Fill 18 lined cups and bake about 20 minutes or until a tester comes out clean.

Cool completely before frosting with either Chocolate Buttercream Frosting (page 168) or Maple Chevre Frosting (below).

{MAPLE CHEVRE FROSTING}

2 STICKS UNSALTED BUTTER
softened

3 OUNCES PLAIN CHEVRE
(page 81) at room temperature

⅔ CUP BROWN SUGAR

¼ TEASPOON SALT

¾ CUP MAPLE SYRUP

¾ TEASPOON VANILLA

1 GENEROUS CUP CONFECTIONER'S SUGAR

Beat the butter and the chevre with the brown sugar and salt at medium speed until light and fluffy. While continuing to beat, add the maple syrup and vanilla. Gradually add the confectioners' sugar until it is a nice fluffy consistency. You may have to add a bit more confectioners' sugar depending on the texture of your chevre.

Part II: The Recipes

MÉMÉRE'S WHOOPIE PIES

> My *mémére* (French-Canadian for grandmother) used to make these and half the neighborhood would be waiting in line for one! She was an amazing cook and loved having me in the kitchen. I have her to thank for teaching me everything I know.

MAKES 15 TO 18 WHOOPIE PIES

FOR THE CAKES:

- 2 CUPS FLOUR
- 1 CUP SUGAR
- 1 EGG
- ⅓ CUP BAKING COCOA
- ⅓ CUP SHORTENING
- 1½ TEASPOONS BAKING SODA
- 1 TEASPOON VANILLA
- 1 CUP GOAT BUTTERMILK *(see 77)*

Preheat the oven to 400 degrees. Place the flour, sugar, egg, cocoa, shortening, baking soda, vanilla, and buttermilk in a mixing bowl and blend well. When the batter is nice and smooth, drop tablespoonfuls on a parchment-paper-lined baking sheet or a really well-greased baking sheet.

Bake for 7 to 10 minutes. Cool on a cooling rack with flat side down. Meanwhile, make the filling.

FOR THE FILLING:

- 4 TABLESPOONS FLOUR
- 1 CUP GOAT BUTTERMILK *(page 77)*
- ½ CUP SHORTENING
- 1 STICK BUTTER *softened*
- 1 CUP SUGAR
- 1 TEASPOON VANILLA

Boil the flour and buttermilk, stirring until it's as thick as paste. Let cool. Cream the shortening, butter, sugar, and vanilla together and add the thickened flour and milk. Beat well until it gains volume and is very fluffy. This takes 8 to 10 minutes.

Spread the flat side of a cake with the filling mixture. Don't be stingy! Top with another cake with the flat side towards the filling. You now have the best whoopie pies you will ever taste!

Cover lightly and keep refrigerated, if you have any left!

OLD-TIME GINGERBREAD

> This is a good old-fashioned gingerbread. It's wonderful served with whipped cream.

SERVES 12 TO 16

1 CUP MOLASSES

½ CUP PACKED BROWN SUGAR

½ CUP BUTTER
softened

2¾ CUPS FLOUR

2½ TEASPOONS GINGER

1 TEASPOON SALT

1 TEASPOON BAKING SODA

½ TEASPOON CINNAMON

½ TEASPOON NUTMEG

½ TEASPOON GROUND CLOVES

½ CUP ORANGE JUICE

¾ CUP GOAT BUTTERMILK
(page 77)

2 EGGS
beaten

WHIPPED CREAM
optional

Preheat the oven to 350 degrees.

Combine the molasses, brown sugar, and butter in a saucepan and heat until the butter is melted. Cool to lukewarm.

In a large bowl, add the flour, ginger, salt, baking soda, cinnamon, nutmeg, and cloves; stir so they are well combined. Add the molasses mixture to the dry ingredients, and then add the orange juice, buttermilk, and the eggs, mixing until smooth.

Pour into a well-greased 13 x 9 x 2 inch pan and bake for 30 to 35 minutes.

Serve with whipped cream.

RHUBARB CAKE

> I love rhubarb in anything. This cake is wonderful with its great rhubarb taste.

SERVES 12 TO 15

1½ CUPS PACKED BROWN SUGAR

1 EGG

½ CUP SHORTENING

2 CUPS FLOUR

1 TEASPOON BAKING SODA

1 TEASPOON VANILLA

1 CUP GOAT BUTTERMILK
(page 77)

2 CUPS CHOPPED RHUBARB

1 TEASPOON CINNAMON

½ CUP SUGAR
brown (can be used if need be)

Preheat the oven to 350 degrees.

Cream together the brown sugar, egg, and shortening. In a separate bowl, sift together the flour and baking soda. Add the vanilla to the milk. Alternately add the flour and buttermilk to the brown sugar mixture. When it's nice and smooth, fold in the rhubarb.

Pour the batter into a greased and floured 13 x 9 inch pan. Mix the sugar and cinnamon together and sprinkle over the batter.

Bake for 30 to 35 minutes.

SUPER-EASY BANANA CAKE

> This is such an incredibly easy cake to make.
> It's big enough to feed a crowd, too!

SERVES 12

1 STICK BUTTER
melted

1½ CUPS SUGAR

2 EGGS
slightly beaten

1 TO 1½ CUPS MASHED RIPE BANANA

2 CUPS FLOUR

1 TEASPOON BAKING SODA

¼ TEASPOON SALT

½ CUP GOAT MILK

GRANULATED SUGAR FOR SPRINKLING
(I use raw sugar for sprinkling because it holds its shape better while baking.)

Preheat the oven to 350 degrees.

Combine the butter, sugar, eggs, and banana. Add the flour, baking soda, and salt. Stir in the milk. You might need to add a little more milk so that you will reach a cake-batter consistency.

Pour into a greased and floured 9 x 13 inch pan.

Bake for 40 minutes. While the cake is still warm, sprinkle evenly with the granulated sugar.

Part II: The Recipes

TRES LECHE CAKE

> This cake is really awesome and is always a big hit! We sometimes serve this with cajeta (below), along with the whipped cream.

SERVES 12

6 EGGS
separated

2 CUPS SUGAR

2 CUPS FLOUR

3 TEASPOONS BAKING POWDER

½ CUP GOAT MILK

2 TEASPOONS VANILLA

ONE (13-OUNCE) CAN EVAPORATED MILK

ONE (14-OUNCE) CAN SWEETENED CONDENSED MILK

ONE (8-OUNCE) CAN CREMA MEDIA
(You can substitute 8 ounces of light cream for this.)

WHIPPED CREAM
optional

Preheat the oven to 350 degrees.

Beat the egg whites until stiff peaks form. Then fold in the sugar gradually to the whites. After all the sugar has been incorporated, add the egg yolks and beat for 3 more minutes.

Combine the flour with the baking powder and fold into the egg mixture alternately with the milk.

Pour into a well-greased 9 x 13 inch pan.

Bake for 35 minutes or until a toothpick comes out clean.

Pour the evaporated milk, sweetened condensed milk, and the crema media into a blender and blend well.

When the cake comes out of the oven, punch holes in it with a toothpick or thin sharp knife. Pour the sauce over the cake, very slowly, letting it get absorbed.

Allow the cake to cool and set completely in the refrigerator.

Serve with whipped cream.

{CAJETA}

¼ TEASPOON BAKING SODA

2 TABLESPOONS CORNSTARCH

3 QUARTS GOAT MILK

3 CUPS SUGAR

Place the baking soda and cornstarch into 1 cup of the milk. Stir until thoroughly dissolved.

Place the rest of the milk and the sugar in a large, heavy-bottom pan. Add the baking soda/cornstarch mixture. Bring to a boil, stirring constantly.

Cook until the mixture is thick and creamy. This can take several hours.

Cool, cover, and refrigerate. This mixture can also be canned!

WALNUT POUND CAKE

Rich, thick, and dense. Oh, so good!

SERVES 12

½ CUP GROUND WALNUTS

2 STICKS BUTTER
softened

2 CUPS SUGAR

3 EGGS

2 TEASPOONS VANILLA

2 CUPS GOAT BUTTERMILK
(page 77)

3 CUPS FLOUR

½ TEASPOON BAKING POWDER

½ TEASPOON BAKING SODA

¾ TEASPOON SALT

Preheat the oven to 350 degrees.

Grind the walnuts in a blender or food processor and set aside. Cream together the butter and sugar, then add the eggs one at a time.

Mix the vanilla into the buttermilk. Sift together the flour, baking powder, baking soda, and salt. Add the nuts, then add the dry ingredients alternately with the buttermilk.

Bake for 1 hour in a well-greased tube pan.

CHOCOLATE WALNUT GOAT MILK FUDGE

The best! Total decadence in each piece.

MAKES 16 SQUARES

3 CUPS SUGAR

1½ CUPS GOAT MILK

⅔ CUPS BAKING COCOA

½ CUP MARGARINE
(I have tried this recipe with butter and for some reason it comes out really dry and crumbly.)

¼ CUP SMOOTH PEANUT BUTTER

1 CUP ROUGHLY CHOPPED WALNUTS

1 TEASPOON VANILLA

In a large pot with a heavy bottom, add the sugar, milk, and cocoa. Stirring often and well, place over high heat to a boil. Continue to boil, stirring often, until the mixture reaches soft ball (240 degrees) stage on a candy thermometer.

Remove from the heat and add the margarine, peanut butter, nuts, and vanilla quickly. Stir until margarine and peanut butter are completely dissolved.

Pour into a well-greased 8 x 8 inch pan. Cool and cut into squares.

PEANUT BUTTER GOAT MILK FUDGE

Everyone's favorite fudge. Rich and creamy and yummy!

MAKES 16 SQUARES

3 CUPS SUGAR

1½ CUPS GOAT MILK

½ CUP MARGARINE
(Substituting butter in this recipe makes for very dry, crumbly fudge!)

1 CUP PEANUT BUTTER
(I prefer super chunky!)

1 TEASPOON VANILLA

In a large pan with a heavy bottom, add the sugar and milk. Place over high heat until boiling, stirring often. Boil until the mixture reaches the soft ball (240 degrees) stage on a candy thermometer.

Remove from heat and add the margarine, peanut butter, and vanilla. Mix well until all the margarine and peanut butter are dissolved.

Pour into a well-greased 8 x 8 inch pan. Cool, then cut into squares.

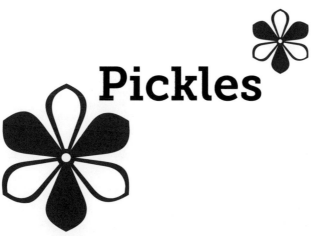

Pickles

Raising goats is an awesome business to be in, however, there are things that must be dealt with, like all of the goat poop that is produced on a daily basis! We compost this "black gold," which has turned our gardens into the most amazing wonderland. There have been times that I wondered if I had wandered into the *Little Shop of Horrors*!

Our cucumbers are prolific, our green beans unrivaled, tomatoes, rhubarb, beets, and so much more need to be used in a good and satisfying way, so why not pickles, relishes, jams, jellies, and salsas?

My grandmother had shared so many of her delicious recipes, which, I received so many compliments from, that I decided to start selling some of the products from these old-time recipes. Now we do hundreds of jars a season and sell them at farmers' markets.

And, the moral of this story is: See! Everything from the goat is usable!

Please remember that these are old-time recipes and jars today and jars back then could vary in size. It will all depend on the size and shape of jars you use as to what your yield will be. I have given you a range of finished product.

> **Disclaimer:** Please follow modern canning instructions. The methods described in these recipes are not intended to be strict instructions. Visit simplycanning.com or your local university extension office for more information.

BREAD AND BUTTER PICKLES

> I can't even describe how good these are! Sweet/sour/crunchy — bet you can't eat just one!

MAKES 5 OR 6 PINTS

8 CUPS SLICED PICKLING CUKES

¼ TO ½ CUP KOSHER OR CANNING SALT PLUS 1 TABLESPOON

2 CUPS WHITE VINEGAR

2 CUPS SUGAR

2 TEASPOONS DRY MUSTARD

2 TEASPOONS TURMERIC

2 TEASPOONS CELERY SEED

2 CUPS SLICED ONIONS

Sprinkle the cucumber slices with ¼ to ½ cup of salt and let them set for 1 hour. Rinse with cold water and drain.

Combine the vinegar, remaining tablespoon of salt, sugar, and spices in a large kettle and heat to boiling. Add the cucumbers and onions, and bring to a boil.

Pack into clean, hot jars leaving ½" head space. Adjust the caps.

Process in a boiling water bath for 5 minutes.

CORN RELISH

Crunchy goodness from the garden.

MAKES 6 PINTS

2 QUARTS CORN
removed from the cob

1 QUART CHOPPED CABBAGE

1 CUP CHOPPED ONION

1 CUP CHOPPED SWEET GREEN PEPPERS

1 CUP CHOPPED SWEET RED PEPPERS

1½ CUPS SUGAR

2 TABLESPOONS DRY MUSTARD

1 TABLESPOON CELERY SEED

1 TABLESPOON MUSTARD SEED

1 TABLESPOON SALT

1 TABLESPOON TURMERIC

1 QUART WHITE VINEGAR

1 CUP WATER

Combine all the ingredients in a large pot, add water, and bring to a boil; reduce heat and simmer for 20 minutes.

Pack the hot relish into clean, hot jars leaving ¼" of head space. Adjust the caps and process 15 minutes in a boiling water bath.

CRISPY GARLIC DILL PICKLES

You can double, triple, or even quadruple this recipe depending on how many cucumbers you want to do!

MAKES 8 TO 10 QUARTS

20 OR MORE PICKLING CUKES
4 CUPS WATER
2 CUPS WHITE VINEGAR
½ CUP KOSHER OR CANNING SALT
½ CUP SUGAR
DILL HEADS*
GARLIC*

* Once you pack the sliced cucumbers into the jars, you will know how many cloves of garlic and dill heads you will need.

Slice the cucumbers lengthwise into spears. Pack into clean, hot jars and add one head of dill and one clove of garlic to each jar.

Boil the water, vinegar, salt, and sugar together and pour into the jars, leaving ½" of head space. Adjust the caps and put into a boiling water bath for 10 minutes.

DILLY BEANS

> Dilly beans are really popular in the New England states. We sell them at the farmers' markets and they go as fast as I can make them!

MAKES 7 PINTS

4 POUNDS GREEN BEANS

1 POUND YELLOW WAX BEANS

1 LARGE CARROT
sliced into thin sticks

6 TABLESPOONS KOSHER OR PICKLING SALT

3 CUPS WHITE VINEGAR

3 CUPS WATER

½ TEASPOON DILL SEED PER PINT
(or 2 fresh dill heads per pint)

½ TEASPOON MUSTARD SEED PER PINT

3 WHOLE PEPPERCORNS PER PINT

1 CLOVE GARLIC PER PINT

Trim the beans and remove strings. Cut the carrot sticks and string beans into lengths that will fit into your jars.

Combine the salt, vinegar, and water, and bring it to a boil.

Add the dill, mustard seed, peppercorns, and garlic cloves to each clean, hot jar. Pack the green beans into the jars and then place the yellow beans and carrots here and there along the outside of the jar so they will show. Cover with the boiling liquid, leaving ½" of head space; adjust the caps. (The beans must be completely covered with the liquid.)

Process in a boiling water bath for 10 minutes.

GARLIC PICKLES

> These pickles are very easy to make and for those who aren't all that enamored with dill, they are a welcomed change.

MAKES 6 QUARTS

2 CUPS WATER

6 CUPS CIDER VINEGAR

3 CUPS SUGAR

1 CUP KOSHER OR PICKLING SALT

1 LARGE ONION
sliced thinly

12 CLOVES GARLIC

15 TO 20 PICKLING CUCUMBERS
sliced lengthwise and cut to fit your jars. (You may need more cucumbers depending on size.)

Place the water, vinegar, sugar, and salt in a pot and heat until boiling.

Place one large piece of garlic in the bottom of each clean, hot jar. Fill each jar with washed and sliced cukes. Put another clove of garlic on the top of the cukes along with a thin slice of onion.

Fill each jar with boiling brine to ½" head space.

Adjust the caps and place the jars in a boiling water bath for 10 minutes.

MUSTARD PICKLES

Another old time favorite. A thickening sauce makes these unusual but really good!

MAKES 4 TO 6 PINTS

2 POUNDS CUCUMBERS
sliced

½ CUP KOSHER OR CANNING SALT

1 POUND ONIONS
sliced

1 LARGE RED BELL PEPPER
chopped

1 QUART APPLE CIDER VINEGAR

9 CUPS SUGAR

1 HOT PEPPER
diced (optional)

3 TABLESPOONS FLOUR

2 TABLESPOONS DRY MUSTARD

3 TABLESPOONS TURMERIC

Cover the cucumbers with water in a non-corrosive pan; add the salt, and let them sit overnight.

In the morning, drain the water and add the onions and pepper. Put in enough cider vinegar just to cover the cucumbers, and then add the sugar. Heat to boiling and boil for 5 minutes.

Mix enough water with the flour, mustard, and turmeric to make a thick liquid. Stir this into the pickles.

Pack into clean, hot jars, and then adjust the caps. Place in a boiling water bath for 5 minutes.

PICCALILLI

> This was one of my dad's favorite relishes. Now it's one of my husband's favorites! Very easy to make and really delicious.

MAKES 6 TO 8 PINTS

6 POUNDS GREEN TOMATOES
chopped

1 LARGE GREEN BELL PEPPER
chopped

1 RED CHILI PEPPER
minced

½ POUND ONIONS
sliced

1 CUP KOSHER OR PICKLING SALT

6 CUPS CIDER VINEGAR

2 CUPS SUGAR

½ TEASPOON GROUND GINGER

½ TEASPOON CINNAMON

1 TABLESPOON MUSTARD SEED

½ CUP HORSERADISH
grated

Combine the tomatoes, peppers, and onions together in a large non-corrosive vessel. Sprinkle with the salt, cover with water, and let stand overnight. Drain completely.

Combine the vinegar, sugar, ginger, cinnamon, and mustard seed, and add this to the vegetables. Simmer until tender, about 15 minutes. Mix in the horseradish.

Pack into hot, clean jars and place in a boiling water bath for 15 minutes.

SOUR PICKLES

These pickles will really make you pucker up!

MAKES 6 OR 7 QUARTS

8 POUNDS PICKLING CUKES
sliced

3 QUARTS CIDER VINEGAR

3 CUPS WATER

¾ CUP KOSHER OR PICKLING SALT

¾ CUP SUGAR

½ CUP MUSTARD SEED

4 PEPPERCORNS PER QUART

Pack the cucumbers into clean, hot jars.

In a large non-corrosive pot, combine the vinegar, water, salt, sugar, and mustard seed, and bring to a boil. Pour the boiling liquid over the cucumbers, leaving about ½" head space. Put the peppercorns in each jar and adjust the lids.

Place jars in a boiling water bath for 5 minutes.

SWEET PICKLE RELISH

> When you are making a really good-quality relish, take the time to grind your vegetables. You can either use a meat grinder (the old-fashioned type that you screw down to the sideboard works really well!) or use your food processor, but don't over-process and turn the veggies into mush.

MAKES 6 PINTS

- 4 CUPS GROUND UNPEELED CUCUMBERS
- 1 CUP GROUND GREEN PEPPERS
- ½ CUP GROUND RED BELL PEPPERS
- 3 CUPS FINELY DICED CELERY
- ¼ CUP KOSHER OR PICKLING SALT
- 3½ CUPS SUGAR
- 2 CUPS WHITE VINEGAR
- 1 TABLESPOON CELERY SEED
- 1 TABLESPOON MUSTARD SEED

Combine the cucumbers, green and red peppers, and celery in a large bowl and sprinkle with salt. Cover with cold water and let it sit for four hours.

Drain thoroughly in a colander and press out all of the liquid. If your colander has particularly large holes, you might want to line it with cheesecloth.

In a large non-corrosive pot, combine the sugar, vinegar, celery seed, and mustard seed, bring to a boil, and keep stirring until the sugar is all dissolved. Simmer for 10 minutes. Add the veggies and return to a boil.

Pack into clean, hot jars leaving about ½" head space. Adjust the caps and place in a boiling water bath for 10 minutes.

ZIPPY ZESTY SALSA

> I use this salsa in many of my recipes! I always make extra to have on hand and to give as gifts. Although this is a zesty-type salsa, it's not overpowering.

MAKES 6 PINTS

10 CUPS CHOPPED ROMA TOMATOES

5 CUPS SEEDED AND CHOPPED GREEN BELL PEPPERS

5 CUPS CHOPPED ONIONS

2½ CUPS SEEDED AND CHOPPED HOT PEPPERS
(A combination of different peppers works great!)

3 CLOVES GARLIC
minced

2 TABLESPOONS FINELY CHOPPED CILANTRO

3 TEASPOONS KOSHER OR PICKLING SALT

1¼ CUPS CIDER VINEGAR

Combine the tomatoes, bell peppers, onions, hot peppers, garlic, cilantro, salt, and vinegar in a large pot and bring to a boil. Reduce the heat and simmer for 10 minutes.

Pack into clean, hot jars and adjust the caps. Place the jars in a boiling water bath for 15 minutes.

ZUCCHINI RELISH

A great way to use up some of that baseball-bat-size zucchini that appeared from nowhere overnight!

MAKES 6 OR 7 PINTS

7 LARGE ZUCCHINI
shredded

6-8 LARGE CARROTS
shredded

2 LARGE ONIONS
shredded

2 LARGE GREEN BELL PEPPERS
shredded

2 TABLESPOONS KOSHER OR PICKLING SALT

2¼ CUPS VINEGAR

¾ CUP BROWN SUGAR

1 TABLESPOON CELERY SEED

1 TEASPOON DRY MUSTARD

Shred the vegetables using a food processor or grater. If you use a food processor, do not over-process!

In a large non-corrosive pot, combine the zucchini, carrots, onions, peppers, salt, vinegar, sugar, celery seed, and dry mustard. Cook about 10 minutes or until the veggies are just beginning to get tender.

Immediately pack into clean, hot jars, leaving about ½" of head space. Adjust the caps and place in a boiling water bath for 20 minutes.

Acknowledgments

Words cannot express the love and thanks that I have for all of the folks who help with *Goat School*! Dawn who travels over an hour and a half each way to help me with the cooking; Tina who makes sure her schedule at the hospital lab will accommodate Goat School so that she can come and help. Tina's children, Lydia and Andrew, who help setting up and with the serving. Stacy, Brady, and especially Martha, who steps in and takes over for me.

Carrie who makes my life so much easier! The day Ken and I met Carrie has been one of the happiest in my life, she milks, she cleans, she makes cheese with me, she's just the greatest!

A special thank you to Andrew Beals from Poulin Grain who comes all the way from northern Vermont to speak about goat nutrition at Goat School. His knowledge, his humor, and his friendship are so appreciated! A very special thank you to Mark of the Brewster Inn in Dexter, Maine, for bugging the newspapers and TV stations about our Goat School and actually getting them interested!

Thank you to illustrator Patrick Corrigan for the unique and funny illustrations and to Chad Hughes for designing the book, and a huge thank you to Kathleen Fleury, my editor, for making my dream come true!

But, most of all, and with the most love you can imagine, thank you Ken. My loving husband without whom life would be so mundane. You are the best!

-Janice

Goat Manual Index

A
angora goats, 21, 62

B
bottle feeding,, 24, 31, 40-42, 45, 59
breeding, 35-37, 69-70
breeds, 20-21
brush clearing, 12, 15
buck record, 70
butterfat, 20-21, 57-58
buying goats, 18-19, 54

C
cashmere, 21, 62
castration, 30-31
coccidiosis, 49
collars, 28

D
dairy goats, 15, 20-21, 56-61
diarrhea, 48-49
doe record, 69
drying off, 61

E
ear tagging, 28

F
feed, 23-27
fencing, 25
fiber goats, 21, 62

G
gestation calendar, 71
goat polio, 49
goat treats, 27
goiter, 48

H
hay, 13, 23-25, 43, 57, 63
hoof scald, 49
hoof trimming, 32-33, 63
horns 31-32

K
kidding 38-45
kidding kit, 39

L
labor, 40-41, 46-47
lice, 48

M
marketing & competition, 52-56
meat goats, 15, 20-21, 63-64
meconium, 42, 44
milking, 15, 57, 59-61
milking stand, 60

P
pet goats, 15
pink eye, 48
placenta, 44

R
record keeping, 37, 68-71
runny noses & colds, 48

S
salt, 24-25
shipping fever, 49
sore mouth, 48

T
tattooing, 29
teats, 48, 57, 59, 61

U
urinary calculi, 49

W
water, 13
worms, 48

Goat Recipe Index

Goat Cheeses
30-Minute Mozzarella, 76
Feta Cheese, 78
Goat Buttermilk, 77
Goat Whole-Milk Ricotta, 79
Hard Cheese or Brick Cheese, 82
Queso Fresco, 80
Stony Knolls Farm Chevre, 81

Salads & Small Bites
Chevre-Stuffed Deviled Eggs, 91
Chevon Taco Salad, 86
Goat Cheese and Honey Dip, 92
Goat Cheese Crostini, 93
Goat Cheese Stuffed Tomatoes, 94
Goat Ricotta and Pasta Salad, 89
Greens, Goat Cheese, and Vinaigrette, 87
Greens with Feta and Croutons, 88
Mozzarella Grilled Mushrooms, 95
Pico de Gallo with Feta, 90
Turkey and Chevre Pinwheels, 96
Zucchini Mini-Quiches, 97
Pie Crust, 97

Big Bites & Main Meals
Cabbage Rolls for Lazy Cooks, 100
Chevon and Tater Tot Delight, 101
Chevon Burgers, 102
Roasted Red Pepper Aioli, 103
Chevon Enchiladas, 104
Chevon Stroganoff, 105
Chevre "Stuffed" Chicken, 106
Dynamites, 107
Easy Homestyle Chevon and Pasta, 112
Goat Sausage Pasta, 108
Goat School Chili, 109
Goat School Mac & Cheese, 110
Honey-Rubbed Chevon Steak, 111
Red Beans and Rice, 113
Sausage and Vegetable Penne, 114
Sautéed Sausage over Angel Hair, 115
Spicy Chorizo Casserole, 116
Stony Knolls Sausage Pizza Casserole, 117
Vegetarian Goat Cheese Lasagna, 118

Breads
Anadama Bread, 122
Banana Nut Bread, 133
Blueberry Lemon Bread, 134
Burger Buns, 123
Buttermilk and Dill Bread, 124
Buttermilk Bread, 125
Cheddar Cheese Bread, 126
Chocolate Chip Pumpkin Bread, 135
Cinnamon (Raisin) Bread, 127
Coconut Bread, 136
Cranberry-Orange Walnut Bread, 137
Date Nut Bread, 138
Garlic French Bread, 128
Honey and Whole-Wheat Bread, 129
Onion Flax Crackers, 141
Peanut Butter Banana Bread, 139
Sally Lunn Bread, 130
Scotch Oatmeal Bread, 131
Seed and Grain Bread, 132
Zucchini Bread, 140

Breakfast
Christmas Morning Sticky Buns, 144-145
Chevon Breakfast Sausage, 146
Sausage Breakfast Casserole, 146
Sour Cream Coffee Cake, 147
Tex-Mex Omelet, 148
Veggie and Chevre Omelet, 149

Cookies
Chocolate Chip Cookies, 152
Chocolate Chip Oatmeal

Cookies, 153

Chocolate Crinkles, 154

Joe Froggers (Big Molasses Cookies), 155

Molasses Coconut Chews, 156

Molasses Gingersnaps, 157

Peanut Butter Cookies, 158

Snickerdoodles, 159

Soft and Chewy Oatmeal Cookies, 160

Sugar Cookies, 161

Cakes & Other Treats

Bread Pudding, 164

Bumpy Apple Cake, 165

Chocolate Walnut Goat Milk Fudge, 176

Chocolate Zucchini Cake, 166

Dark Chocolate Buttermilk Cake, 167

    Amazing Frosting, 167

Goat School Cherry Cupcakes, 168
    Chocolate Buttercream Frosting, 168

Maple Cupcakes, 169
    Maple Chevre Frosting, 169

Memere's Whoopie Pies, 170

Old-Time Gingerbread, 171

Peanut Butter Goat Milk Fudge, 177

Rhubarb Cake, 172

Super-Easy Banana Cake, 173

Tres Leche Cake, 174
    Cajeta, 174

Walnut Pound Cake, 175

Pickles

Bread and Butter Pickles, 180

Corn Relish, 181

Crispy Garlic Dill Pickles, 182

Dilly Beans, 183

Garlic Pickles, 184

Mustard Pickles, 185

Piccalilli, 186

Sour Pickles, 187

Sweet Pickle Relish, 188

Zippy Zesty Salsa, 189

Zucchini Relish, 190

Notes

Notes

Notes